Environmental Solutions for Grounds Management

(Nationally Awarded Program)

Kevin Scott Mercer, CGM, CSFM

About The Author

I remember working in the garden or mowing and trimming around my grandparents' house alongside my grandfather, who was teaching me about attention to detail at the early age of nine. He taught me so much and always told me to be proud of my work and never leave a job unfinished. I remember being in my early twenties, in and out of jobs, when I applied for a groundskeeper job on a nine-hole golf course in Gettysburg, Pennsylvania, in March 1995 as summer help. During the first month on the job, I knew this was what I wanted to do for the rest of my life. After taking some college courses in landscape management, I landed my first assistant golf course superintendent job at Andrews Air Force Base. A few years later, I earned my first grounds manager job at St Mary's College of Maryland for almost ten years. Then, I moved to New York, where I accepted a job at Vassar College in Poughkeepsie. After a few years, I eventually found what I was looking for at Denison University and found the best grounds team in the country.

When I look back at my career and all the years working in the liberal arts arena, I have to admit the students have taught me so much about our planted earth and all the environmental issues we face. I want to thank all of them and give them a shoutout. To this day, I still focus my ground management program on stormwater runoff, protecting wetlands, promoting wildlife, native plantings, and pollinators, and reducing greenhouse gas.

Through my years of professional grounds management, I've found my strategic philosophy has been to stay the course for business but implement a strategy that showcases our environmental stewardship programs while providing a return on investments without altering the aesthetics and function of the landscape. I have managed golf courses, sports fields, and college and university grounds.

Inclusion and transparency are cornerstones for successfully adopting new strategies that result in more sustainable practices with acceptable returns on investment and utilizing more environmentally friendly practices (if feasible). I have learned the best approach to building relationships with team members and the community is fostering relationships with follow-through communication and your actions by being yourself with concern for others.

I want to thank you for purchasing my book and for being great environmental stewards. I started as a grounds manager in 2004 at St Mary's College of Maryland, and to date, I have won six prestigious national environmental awards from the Professional Grounds Manager Society (PGMS) and the Sports Fields Manager Association (SFMA).

I am so happy to share with you what have I have learned over the years on things that work from my own experiences. I want to be honest and share with you that I do not have a PhD, master's or even a bachelor's degree, I have something far more worthy, which is over thirty years of failures and successes in the grounds maintenance profession.

Nature has provided us with clean air, fresh water, and soil to showcase our landscape and professionalism. Over the last fifty years, rigorous environmental regulations and laws have been put in place to protect our local watersheds, air, wildlife habitats, pollinators etc. My intention for this book is intended to help ground managers, horticulturists, landscape companies, golf course superintendents, governmental agencies and educational institutions with practical solutions for today's challenging environmental issues.

The purpose of the book is to give you a proactive approach for today's environmental concerns. My goal is not to take away your pesticides or fertilizers from maintenance programs, but to strengthen your existing landscape with solutions that can help reduce stormwater runoff and add biodiversity for insects, wildlife and plants. We all should have a social responsibility to protect our environment and to promote awareness through outreach and education. The land we maintain is only borrowed before we hand it off to the next generation.

I would like to dedicate this book to my grandfather, mother, father, wife, and children who have supported me throughout my career along with some of the best mentors in the business. I also want to thank the Professional Grounds Managers Society (PGMS) and the Sports Fields Managers Association (SFMA) for their years of professional development. A special thanks to Jessica Hanvey for all the wonderful artwork illustrations.

Kevin Scott Mercer CGM, CSFM

Kevin S Mercer

Book Endorsements

"Environmental Solutions for Grounds Management" by Kevin Mercer is a fantastic resource for any grounds manager looking to integrate sustainable practices without compromising aesthetics. Drawing from years of professional experience, Kevin provides practical strategies that showcase environmental stewardship while ensuring a good return on investment. This book is a must-read for anyone looking to make their grounds management practices more environmentally friendly.

Jeff McManus

Director of Landscape and Waster Services

Ole Miss

Sustainability is the use and reuse of resources to leave the environment in a usable maintainable capacity for our future generations. Kevin has combined his learning and the school of hard knocks to put together this collection of ideas and definitions for a quick read and reference. The information shared is good for both new and veteran greens industry practitioners.

Roger Conner, CGM

Duke University Grounds Superintendent, Sr.

Kevin Mercer's book Environmental Solutions for Grounds Management results from several decades of work to enhance outdoor environments ranging from golf courses to college campuses. I had the privilege of working with Kevin and his team at St. Mary's College of Maryland, where it quickly became apparent to me that Kevin is dedicated to both his team of coworkers and the outdoor surroundings that all enjoy. His work, clearly explained and illustrated in Environmental Solutions, consistently looks to give back sustainably.

His sections on enhancing soil and compost, encouraging biodiversity, and managing the grounds are clear, engaging, and supported by extensive research and years of practical experience. I highly recommend Environmental Solutions for Grounds Management to experienced grounds managers and those who are interested in learning more about creating a beautiful and welcoming outdoor presence responsibly.

Dr. Thomas J. Botzman (Retired)

Former President of Misericordia University and University of Mount Union

"Kevin highlights best practices for environmental and grounds care for everyone; from the average home owner to the professional grounds manager. From his years of experience, he shares how you can better environmental experiences for generations to come, by using safe practices and treating nature with a key ingredient: nature."

Chris Mason

Director, Field Operations

Groundskeeper Reflection

discipline acts of pride, subjection, and fortitude

cultivates a deeper indebtedness from mother nature attributes.

a groundskeeper is an unknown artist, balancing forms of emotion, science, and mediation

awaiting the cold winter dismissal and preparing the grounds canvas for its new spring painting

from an array of herbaceous colors with precise and gentle strokes of passion

to the rolling hills of the green grasses that we all interned from Scotland

as the west winds bring us wisdom and it also brings with it sorrows from lessons of failures

that gives us humble experiences as we grow into our persona

from an appreciation that a few processes and only fewer apprehend through trials of disappointment and gratitude

we find different prospectives and individual disciplines that direct us to our perfection

comprehension is like a weeping willow gently blowing her poetic leaves in the warm spring air

impressionable sunrise of thick morning dew on the fresh-cut grass leaves us with its intoxicating smell that awakens our senses

the quiet eye of a groundskeeper is an architect of peaceful solitude

like a maestro directing a sympathy of colors or giving songbirds a place of shelter

Frederick Law Olmsted echoes within our souls from his teachings

that we learned from to honing-in our unique charismatic styles that reflect our passion

Kevin Scott Mercer

Table of Contents

Section 2: Stormwater Management Introduction

Stormwater is the water that flows from land after it rains and melting snow. If not appropriately managed, it can pick up pollutants as they flow and cause flooding. Stormwater management is the practice of controlling stormwater flow to protect water quality and prevent flooding by slowing down the water and having it go through a filtering process using native vegetation. Stormwater runoff can impact our day-to-day life in both positive and damaging ways. Some negative effects could be flooding, stormwater nutrients contaminating local watersheds, and erosion, leading to problems like landslides and sedimentation in waterways, causing property damage that is harmful to aquatic life. Stormwater can be beneficial, too; for instance, it can replenish groundwater supplies, especially for those living in the county's water-shortage areas. Properly constructed rain gardens and bioswales can create valuable habitats for plants and animals. There are several diverse ways to manage stormwater, including:

- Green infrastructure: This includes things like rain gardens, bioswales, and permeable pavements. These features allow stormwater to soak into the ground, which helps to filter out pollutants and reduce flooding.

- Gray infrastructure: This includes things like storm drains, pipes, and detention ponds. These features collect and store stormwater, which can help to prevent flooding.

- Best management practices (BMPs): These are specific actions that can be taken to reduce stormwater pollution. Examples of BMPs include picking up pet waste, using pesticides and fertilizers sparingly, and sweeping up driveways and sidewalks.

- Stormwater management protects water quality, prevents flooding, and ensures a sustainable water supply.

Why is stormwater management important? According to the Environmental Protection Agency (EPA), impervious surfaces, such as pavement and roofs, prevent precipitation from soaking into the ground in established areas. Instead, water runs rapidly into storm drains, sewer systems, and drainage ditches, picking up pollutants like trash, chemicals, oils, and dirt or sediment along the way that can harm rivers, streams, lakes, and coastal waters while causing serious destruction like:

- Downstream flooding

- Stream bank erosion

- Increased turbidity (muddiness created by stirred-up sediment) from erosion

- Habitat destruction

- Combined storm and sanitary sewer system overflows

- Infrastructure damage

- Contaminated streams, rivers, and coastal water

Citation

- US EPA, OARM. "EPA Facility Stormwater Management | US EPA." US EPA, Mar. 2019, www.epa.gov/greeningepa/epa-facility-stormwater-management. Available at https://www.epa.gov/greeningepa/epa-facility-stormwater-management

2.1: Stormwater Dialect

Stormwater professionals might use specific terms and acronyms to communicate efficiently. This "dialect" may include terminology like BMPs (Best Management Practices), SWMM (Storm Water Management Model), NPDES (National Pollutant Discharge Elimination System), or infiltration rates. To better understand stormwater management, knowing and understanding the acronyms and meaning is imperative.

1. **Best Management Practices (BMPs)**

Stormwater controls are used by communities, construction companies, industries, agriculture, government agencies, and others to filter out stormwater pollutants and/or prevent pollution by controlling it at its source.

2. **Nonpoint Source (NPS)**

Stormwater pollution is derived from events causing sheet-water movement from waterlogged lawns and non-pervious surfaces that picks up everything in its path (i.e., fertilizers, pesticides, oil, animal manure, and sediment).

3. **Point Source**

Stormwater pollution comes from a single point (i.e., sewage treatment plants, coal-burning plants, factories discharging wastewater, oil spills, etc.).

4. **Total Maximum Daily Load**

Total Maximum Daily Load (TMDL) calculates the highest amount of a pollutant that a body of water can take in and still meets the standards for healthy ecological systems.

5. Erosion and Sediment Control (ESC)

Temporary construction or flash flooding along non-planted streams or riverbanks can cause disturbance in the soil, that can cause soil erosion and sediment buildup in watersheds. This is a practice of preventing or reducing sediment movement from a site during construction by implementing man-made structures, land-management techniques, or natural processes like silt fences, water retention areas, and rain gardens.

6. Clean Water Act

The Clean Water Act (CWA) has focus points to prevent, reduce, and eliminate pollution in all of the eighteen water basins outlined by the U.S. watersheds with a direction: "restore and maintain the chemical, physical, and biological integrity of" the eighteen river basins.

7. Stormwater Pollution Prevention Plan

A Stormwater Pollution Prevention Plan (SWPPP) is a site-specific written document that outlines all of the activities and conditions at a site that could cause water pollution and details the steps the facility will take to prevent the discharge of any stormwater pollution.

8. National Pollutant Discharge Elimination System

The National Pollutant Discharge Elimination System (NPDES) requires industries that are point sources of stormwater nutrients and pollution to hold permits before pollutants can be discharged into navigable waters. These sources must also maintain records and monitor pollutant discharges daily, weekly, monthly, or yearly.

Citation

"Glossary of Terms & Acronyms - Public Works - Stanislaus County." Stanislaus County, www.stancounty.com/publicworks/storm/glossary.shtm

Halfacre, A. C., Hitchcock, D. R., & Shuler, J. A. (2007). *Community Associations and Stormwater Management. Community Associations and Stormwater Management: A Coastal South Carolina Perspective*, 7-20,35,36.

Figure 1: This is a poorly installed silt fence allowing stormwater sediment and nutrients to enter into the local watershed during a spring thunderstorm.

Figure 2: Frequent use of sports fields can lead to compacted soil. This reduces the infiltration rate, meaning less water soaks into the ground, and more ends up as stormwater runoff, taking fertilizer nutrients with it.

2.2: U.S. River Basins

The United States has a complex network of rivers that drain into various bodies of water, including the Atlantic Ocean, Pacific Ocean, Gulf of Mexico, and Arctic Ocean. These rivers are grouped into major drainage basins based on their water flow direction.

The United States boasts a diverse and vibrant network of over 250,000 rivers, carving its path through the landscape and shaping the nation's geography, ecology, and water resources. These rivers aren't isolated entities; they flow within distinct drainage basins, interconnected systems that collect and channel water toward larger bodies like oceans, lakes, or underground aquifers. Understanding these basins is crucial for managing our precious water resources effectively. The contiguous United States can be broadly divided into 18 major river basins, each with unique characteristics and challenges. The longest river in the United States is the Missouri River, which is 2,450 miles long. Nearly all U.S. rivers provide drinking water, irrigation water, transportation, electrical power, drainage, food, and recreation. We depend on rivers for survival and overall quality of life. All the rivers in the contiguous states feed into the eighteen river basins listed in Table 1. The phrase Best Management Practices, or BMPs, was created more than thirty years ago to describe acceptable practices to protect our local river basins. There are eighteen major rivers/basins in the lower forty-eight U.S. states, as designated by the U.S. Water Resources Council (see Table 1). One inch of rain falling on one acre of ground is equal to about 27,154 gallons of water and weighs about 113 tons. An inch of snow falling evenly on one acre of ground is equivalent to about 2,715 gallons of water. However, this amount of stormwater can cause serious consequences. These basins are more than just geographical features; they're the lifeblood of communities, ecosystems, and industries. Understanding their complex dynamics and the challenges water basins face is vital for ensuring sustainable and clean tributaries for the United States water basins. Here's a list of the major U.S. river basins:

Table 1: U.S. River Basin Chart

	Basin Name	U.S State(s)
1	Pacific Northwest Basin	Washington, Oregon, and Idaho
2	California River Basin	California
3	Great Basin	Nevada and Utah
4	Lower Colorado River Basin	Colorado and Arizona
5	Upper Colorado River Basin	Colorado
6	Rio Grande River Basin	New Mexico and Texas
7	Texas Gulf Basin	Texas
8	Arkansas White Red Basin	Oklahoma and Texas
9	Lower Mississippi River Basin	Arkansas and Louisiana
10	Missouri River Basin	Montana, Nebraska, South Dakota, North Dakota, Wyoming, Missouri and Kansas
11	Souris-Red-Rainey Basin	Minnesota
12	Upper Mississippi River Basin	Wisconsin and Illinois
13	Great Lake Basin	Michigan, New York, and Pennsylvania
14	Tennessee River Basin	Tennessee
15	Ohio River Basin	Ohio, Kentucky, West Virginia, Western Pennsylvania, and Indiana
16	South Atlantic Gulf Basin	North Carolina, South Carolina, Georgia, Florida, Mississippi, and Alabama
17	Mid Atlantic Basin	New York, Pennsylvania, Maryland, Delaware, Connecticut, New Jersey, and Virginia
18	New England Basin	Maine, New Hampshire, Massachusetts, and Rhode Island

Source: Contiguous U.S. Major River Basins as designated by the U.S. Water Resources Council

Citations

NCEI.Monitoring.Info@noaa.gov. "Geographical Reference Maps | National Centers for Environmental Information (NCEI)." Geographical Reference Maps | National Centers for Environmental Information (NCEI), Www.ncdc.noaa.gov, https://www.ncdc.noaa.gov/monitoring-references/maps/us-river-basins.

MACDONALD, CHEYENNE, and MARK PRIGG. *The Veins of America: Stunning Map Shows Every River Basin in the US*, October 21, 2016, https://doi.org/https://www.reeldealanglers.com/the-veins-of-america-stunning-map-shows-every-river-basin-in-the-us/.

2.3: Stormwater Nutrients

Two of the most problematic stormwater nutrients are nitrogen and phosphorus. Stormwater runoff of nutrients from developed landscapes is recognized as a major threat to water quality degradation through cultural eutrophication or algae bloom, which can lead to poor water quality, cause major imbalances and harmful algae growth, and have a catastrophic effect on aquatic wildlife. The major contributors to these issues are fertilizers, nitrous oxide/sulfur dioxide, sewage, animal manure, sediment, oils/grease, and detergents.

1. **Fertilizers:** Fertilizers' macronutrients—nitrogen (N), phosphorus (P), and potassium (K)—are regularly used in commercial and noncommercial turfgrass applications. According to the U.S. EPA's New England Regional Laboratory, 40 to 60 percent of the nitrogen that people put on their lawns through fertilizer winds up in "surface and groundwater. These nutrients are essential for proper turfgrass health and durability. A high rate of water-soluble nitrogen (WSN) will leach more quickly into stormwater runoff when compared to water-insoluble nitrogen (WIN), which is considered a nonpoint source (NPS).

2. **Nitrous Oxide/Sulfur Dioxide:** These nutrients dissolve easily in water and can be carried very far by the wind at high altitudes. As a result, the compounds can be transported in the form of rain, sleet, snow, and fog, commonly called acid rain. When acid rain falls, it raises the nutrient levels in local river basins. Acid rain is a nonpoint source. Coal-burning power plants, the electric power industry, gasoline-powered transportation, and agriculture have increased the amount of nitrogen in the air by using fossil fuels.

3. **Sewage:** Sewage and septic systems are responsible for treating large quantities of waste, and these systems do not continuously operate properly or remove enough nitrogen and phosphorus before discharging into local water basins. Coliform presence in stormwater indicates fecal contamination from humans and animals. Coliform cause diarrhea and other dysenteric symptoms once ingested. When levels are high, this could also shut down beaches and hurt the local economy from tourism.

4. **Animal Manure:** Animal manure from raising livestock over the last century has rapidly changed our ecosystem and contributed to a sharp increase in nutrient levels in local water basins. Animal production is increasing, and as a result, additional building is occurring further away from feedstock supplies, making it harder to spread manure. The large quantity of manure these operations produce is applied to land as fertilizer, stacked in feedlots, or stored in lagoons. An oversupply of manure frequently means that it is applied to crops in excess, even in the winter months when temperatures are freezing, further exacerbating nutrient runoff and leaching.

5. **Sediment:** Sediment is loose sand, clay, silt, and other soil particles that settle at the bottom of a body of water. Sediment can come from soil erosion or the decomposition of plants and animals. Sediment is transported by wind, water, and ice to streams, lakes, and rivers. Sediment is a nonpoint source.

6. **Oil and Grease:** Oil and grease can prevent oxygen from entering surface water and create a chemical oxygen demand that consumes dissolved oxygen, thereby stressing aquatic systems.

7. **Detergents:** The breakdown of phosphorus in detergent can contribute to an oversupply of phosphate in waterways and cause an imbalance of the aquatic ecosystem from eutrophication.

Fox, Radhika, et al. Addressing Nutrient Pollution in Our Nation's Waters, March 23, 2017, pp. 1–6.

Citations

Mooney, C. (2015, March 11). Americans are judging their neighbors' lawns — with surprising environmental consequences - the Washington Post. Americans are judging their neighbors' lawns — with surprising environmental consequences. https://www.washingtonpost.com/news/energy-environment/wp/2015/03/11/forget-what-your-neighbors-think-stop-dousing-your-lawn-with-so-much-fertilizer/

Table 2: Examples of Nonpoint Source and Point Source Stormwater Nutrients

Point Source / Nonpoint Source	Urban Landscape Issues	The Problem
Sediment / Nonpoint Source	Sediment, dirt, and sand on roads, driveways, and parking lots or eroded sediment from disturbed surfaces (e.g., construction sites) enter streams with stormwater runoff; increased flow causes stream bank erosion.	Carries pollutants, erodes stream channels and banks, and destroys in-stream habitats
Nutrients / Nonpoint Source	Excess fertilizers on lawns or fields, failing septic systems, and animal waste	Stimulates excessive plant growth, lowers dissolved oxygen levels, degrades aesthetics and destroys native aquatic life
Hot Surfaces / Point Source	Warmer water is caused by runoff from impervious surfaces, removal of streamside vegetation, and reduction in groundwater flows.	Harmful to cold water-species aquatic life, promotes the spread of invasive species and excessive plant growth, and reduces dissolved oxygen levels in water.
Bacteria / Coliform Nonpoint Source/ Point Source	Bacteria: potentially pathogenic microscopic organisms in failing septic systems, sewer overflows, and animal (including pet) waste	Harmful to humans; untreated waste can cause numerous diseases
Chemicals / Nonpoint Source	Toxic contaminants and heavy metals such as mercury, cleaning compounds, pesticides, and herbicides; industrial by-products such as dioxin; and vehicle leakage of oil, gas, etc.	Harmful to humans and aquatic life at fairly low levels; many resist break down, and some accumulate in fish and other animal tissues (including human) and can lead to mutations, disease, or cancer.

Source: https://www.epa.gov/nps/nonpoint-source-urban-areas

"Nonpoint Source: Urban Areas | US EPA." US EPA, Www.epa.gov, September 15. 2015, https://www.epa.gov/nps/nonpoint-source-urban-areas.

2.4: Eutrophication

Eutrophication occurs when the excessive richness of nutrients in a lake or other body of water, frequently due to runoff from the land, causes a dense growth of plant life and death of aquatic life from lack of oxygen. Nitrogen + phosphorus + phytoplankton = algae bloom, also known as eutrophication, which causes red tide in salinity waters and green/blue tide in freshwater lakes and rivers. They both result in immense fish kills.

Citations

Malone, Thomas C., and Alice Newton. "The Globalization Of Cultural Eutrophication In the Coastal Ocean: Causes And Consequences." Frontiers, Www.frontiersin.org, 1 January. 2001, https://www.frontiersin.org/articles/10.3389/fmars.2020.00670/full.

"Harmful Algal Blooms: Red Tide Vs. Blue-green Algae." UF/IFAS Extension Osceola County, Blogs.ifas.ufl.edu, August 8. 2018, https://blogs.ifas.ufl.edu/osceolaco/2018/08/08/harmful-algal-blooms/.

Figure 3: Eutrophication is the process by which a body of water is overloaded with nutrients like fertilizers, sewage, and runoff from commercial and residential areas. It depletes dissolved oxygen in the water, which starts the algea bloom process.

Figure 4: Show here is a Denison University employee using a dissolved oxygen water meter to test our irrigation pond dissolved oxygen, temperature, and pH

Figure 5: "A healthy pond is a balanced ecosystem. Regular water testing is crucial for maintaining this delicate equilibrium. You can identify potential problems early by monitoring key parameters and taking corrective actions to protect your pond's inhabitants. The water test should include a pH test between 6.5-8.5. The temperature should be around 68 to 74 degrees Fahrenheit. Your dissolved oxygen (DO) should read around 7.5 and 8.5 ppm. Ensure your ammonia, nitrite, and nitrate levels are low and not toxic".

2.5: Prevention Methods

Stormwater runoff can cause several environmental problems, from rapidly moving water that could erode stream or ditch-line banks to altering or having catastrophic effects on aquatic wildlife. Stormwater runoff can push excess sediment into rivers and streams, blocking the sunlight needed for photosynthesis from reaching beneficial filtering plants and suffocating bottom feeders that help filter water for ecological and biological control from lack of oxygen. Below is a list of five examples that can be added to strengthen a BMP program.

1. **Fertilizers**: All fertilizers have two ways of breaking down the coating and releasing nutrients for the plant: water-soluble nitrogen (WSN) and water-insoluble nitrogen (WIN).

- Inorganic water-soluble nitrogen (WSN) and synthetic organic urea are released quickly into the soil, which can increase the risk of leaching at high rates into stormwater. Inorganic sources include ammonium nitrate, ammonium sulfate, potassium nitrate, calcium nitrate, and mono-di-ammonium phosphate.

- Water-insoluble nitrogen (WIN) is a slow-release nitrogen, listed on the fertilizer bag as water- insoluble nitrogen (WIN). If no WIN is listed on the fertilizer label, assume that all the nitrogen is water-soluble or fast-release nitrogen. Do not apply fertilizer before a rain event. If fertilizers are overthrown on non-pervious surfaces such as roadways, parking lots, and sidewalks, they will be picked up and carried by rain events and stormwater. Sweep or blow excess fertilizer spills on non-target areas back into the lawn or ornamental beds. Get a soil test to see what your soil needs. Fertilizers with slow-release water-insoluble nitrogen (WIN) are safer for the environment. When applying slow-release fertilizer, a good habit is to cover all drains in lawn areas to avoid any runoff from rain events.

2. **Pesticides**: Pesticide applications should always be conducted as part of an integrated pest management (IPM) plan. A well-defined IPM plan should be based on prevention, monitoring, and control, which offer the opportunity to eliminate or drastically reduce the use of pesticides and minimize the toxicity of and exposure to any used products. Always store pesticides, used oil, antifreeze, and other chemicals on secondary containment pallets. In addition, ensure spill kits are available anywhere chemicals are stored and near fuel pumps.

3. **Erosion**: Erosion protection can be executed by protecting soil through planting groundcover vegetation or native trees and shrubs. Soil washed away by rain can pollute streams and lakes. When renovating an athletic field or lawn area of one acre or more, consider using a slit fence or bales of straw around the perimeter to avoid erosion and lost soil. Plan to renovate in months not prone to thunderstorms and heavy rains. Reduce soil erosion by planting native groundcovers on exposed soil, such as under trees or on steep slopes.

4. **Storm Drains / Secondary Containment**: Keeping storm drains clean and debris-free will help reduce sediment entering the local water basin.

5. **Buffer Strips / Filtering:** Buffering strips serve two purposes: filtering stormwater pollutants and providing food and shelter for wildlife.

2.6: Stormwater Buffer Filtration Systems

In large, open areas such as parking lots, athletic fields, large open lawn areas, and grassy hills landscapes could cause stormwater to flow rapidly which is called sheet water movement during heavy downpour rain events. Slowing down stormwater movement from downpours from rain events is crucial to prevent algae blooms and rising temperatures from entering your local estuaries. In (Figure 3) listed below are three colors that represent a functional buffer zone to help aid greatly in slowing down stormwater.

Citation

"Riparian Buffers: Using the Power Of Plants To Help Clean Our Waterways." Penn State Extension, Extension.psu.edu, February 14. 2020, https://extension.psu.edu/riparian-buffers-using-the-power-of-plants-to-help-clean-our-waterways.

Holm, Bobbi A, et al. "Stormwater Management: What Stormwater Is and Why It Is Important ." July 2014, pp. 1–4. https://doi.org/https://extensionpublications.unl.edu/assets/pdf/g2238.pdf.

Figure 6: shows a diagram of how to install a stormwater buffer around a pond, lake, stream, etc.

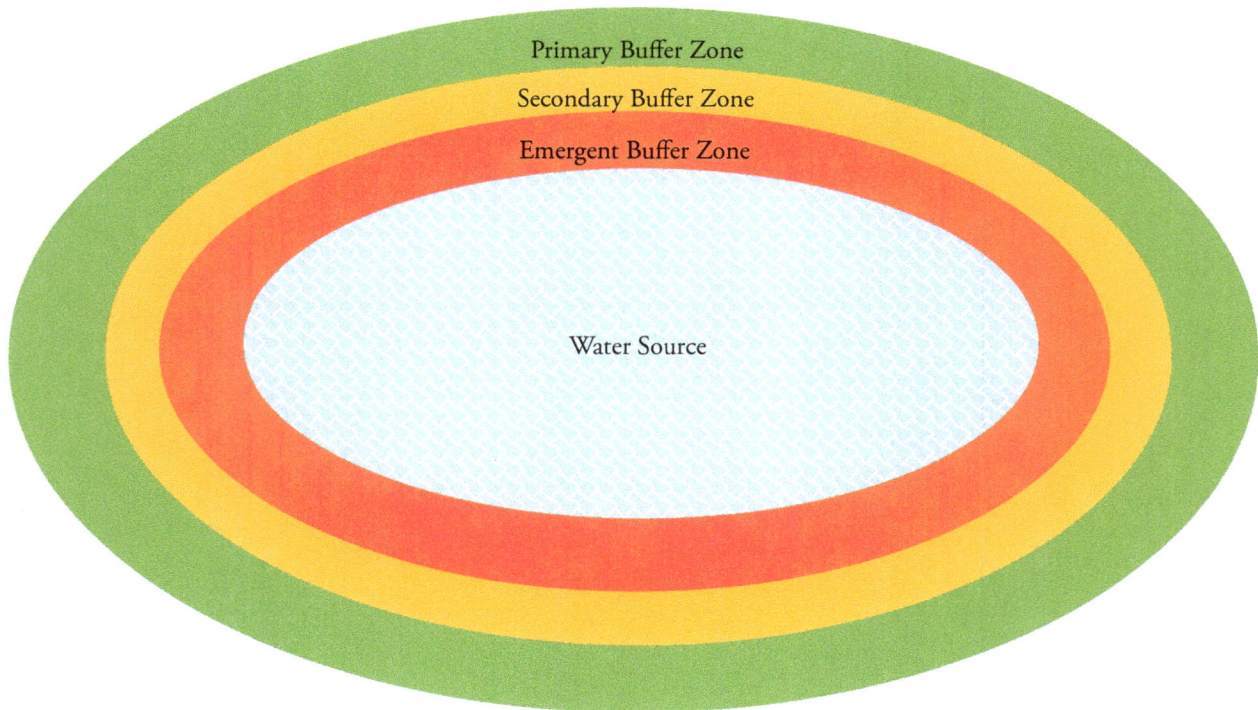

The red area represents the riparian/ emergent buffer zone. Emergent plants are a vital component in stormwater management. They are rooted in the soil along the shoreline but have leaves that emerge above the water's surface. These plants play a crucial role in improving water quality by filtering pollutants, reducing erosion, and providing habitat for wildlife. This zone should be 5 to 10 feet wide from the shoreline.

The yellow area represents the secondary buffer zone. This area is made of native trees and shrubs. Native plants are ideal for stormwater management due to their deep root systems, which help to increase infiltration, reduce erosion, and filter pollutants from stormwater runoff. This area should be 10 to 25 feet wide.

The green area represents the primary buffer zone. Native groundcovers and grasses are essential components of a healthy stormwater buffer. Their role is to slow the water down, filter pollutants, prevent erosion, and increase infiltration is invaluable. This area should be 15 to 25 feet wide.

"Riparian Buffers: Using the Power Of Plants To Help Clean Our Waterways." Penn State Extension, Extension.psu.edu, February 14. 2020, https://extension.psu.edu/riparian-buffers-using-the-power-of-plants-to-help-clean-our-waterways.

2.7: Buffer Management Strips

The Latin word "riparian zone" refers to areas housing rivers, lakes, and wetland banks. These areas are a vital part of our ecological system. Riparian buffers are vegetated areas of native perennials, shrubs, and trees. Buffer strips slow down water runoff and trap sediment and infiltration within the buffer strip. Buffers also trap fertilizers, pesticides, pathogens, and heavy metals, and they help trap snow and cut down on blowing soil in areas with strong winds. In addition, they protect livestock and wildlife from harsh weather and buildings from wind damage. If properly installed and maintained, they have the capacity to remove up to:

*   50 percent or more of nutrients and pesticides

*   60 percent or more of certain pathogens

*   75 percent or more of the sediment

There is little or no cost involved in protecting existing riparian buffers. Existing buffers can be protected through ordinance requirements, easement agreements, or conscious efforts to decrease mowing and maintain trees and shrubs. It is a social responsibility to preserve and improve riparian buffers. Over the last several decades, the amount of urban and suburban forest trees processed into timberlands has been substantial. Each year, thousands of acres of prime forest are lost. While timber harvest influences watershed functions, the effects are relatively short-term compared to the impacts of conversion and urbanization. The changes associated with development are more significant and permanent. The rate of conversion of forestland to developed uses is accelerating. The impacts of watershed urbanization have been accelerating as well. They include increased stormwater runoff, accelerated erosion, sedimentation, degraded water supplies, and severe disturbance of aquatic and marine systems. In fact, recently, several types of salamanders, frogs, and fish species have been added to the U.S. Fish and Wildlife Endangered list due to watershed urbanization.

Native plants are trees, shrubs, flowers, grasses, and groundcover that occur naturally in the five climate zones of the United States. Native plants can adapt to wet or dry areas, hot or cold. Another great feature about native plants is that native wildlife depend on them for shelter and food. Native plants do not just slow stormwater to prevent erosion. They also serve as natural biological filters. Native plants are resistant to most local pests and diseases. In regions with heavy clay soil, deep-rooted native plants can break up the soil better than typical lawn grass varieties and improve clay soil's permeability, ultimately acting as a green stormwater alternative. There is greater infiltration from native plants, which results in better pollutant filtering and more water replenishment of the aquifer, which reduces flood water and stormwater impacts.

In summary, utilizing native plants and trees in landscaping allows for short- and long-term positive effects on stormwater runoff quantity and quality. Rebuilding our national ecological system will take time and resources, but together, we can positively impact the next generations to manage and strengthen. Table 3 is a key chart that can help choose plants for particular light conditions. Green key charts represent the primary zone for planting. Yellow represents the secondary planting zone. Red represents the emergent (riparian) planting zone.

Citation

"SL433/SS647: What Else Can Surface Water Buffer Systems Do?—Exploring Multiple Ecosystem Services." SL433/SS647: What Else Can Surface Water Buffer Systems Do?—Exploring Multiple Ecosystem Services, Edis.ifas.ufl.edu, https://edis.ifas.ufl.edu/publication/SS647.

Pryor, Randy, et al. Buffer Strip Plant Selection, October 31, 2019Conservation Buffers." Megamanual. geosyntec.com, https://megamanual.geosyntec.com/npsmanual/conservationbuffers.aspx.

Table 3: Key Chart for Buffering Plants

☀	Plants that require full sun.
⛅	Plants that require both shade and sun.
☁	Plants that require very little sun.
♺	Plants that do a great job filtering many types of stormwater nutrients.

Table 4: Primary Buffer Strip Native Plant List

A primary buffer strip has one function: to slow rapidly flowing water from rain events. The primary buffer strip contains tightly planted native groundcovers and grasses to act as barriers and absorb moving water during rain. Keeping the buffer strip native will aid in wildlife biodiversity as well. When selecting native plants for a primary buffer strip, check the local and state invasive species list so as not to encourage the proliferation of invasive plants that threaten native plants and wildlife.

Common Name	Scientific name	☀	⛅	☁	USDA Planting Zone
Maidenhair Fern	*Adiantum pedatum*			X	2-9
Canada windflower	*Anemone canadensis*	X	X		2-9
Red Bearberry	*Arctostaphylos uva-ursi*	X	X		3-8
Goat's Beard	*Aruncus dioicus*	X	X		4-8
Large Leaf Aster	*Eurybia macrophylla*		X	X	3-8
Lady Fern	*Athyrium filix-femina*		X	X	4-8
Bristle Leaf Sedge	*Carex eburnea*	X	X		2-8
Palm Sedge	*Carex muskingumensis*	X		X	2-8
Pennsylvania Sedge	*Carex pensylvanica*	X	X	X	3-8
Blue Mist Flower	*Conoclinium coelestinum*	X	X		4-9
Tickseed	*Coreopsis rosea*	X			3-8
Hayscented Fern	*Dennstaedtia punctilobula*	X	X	X	3-8
Wild Strawberry	*Fragaria virginiana*	X	X		3-9
Wild Geranium	*Geranium maculatum*	X	X		3-9
Prairie Smoke	*Geum triflorum*	X			3-9
Crested Iris	*Iris cristata*	X	X		4-9
Feathery False Solomon's Seal	*Maianthemum racemosum*		X	X	3-8
Sensitive Fern	*Onoclea sensibilis*	X	X	X	3-9
Cinnamon Fern	*Osmunda cinnamomea*	X	X	X	3-10
Royal Fern	*Osmunda regalis*	X	X		3-9
Running Groundsel	*Packera obovata*	X	X		4-8
Obedient Plant	*Physostegia virginiana*	X	X		4-9
Woodland Stonecrop	*Sedum ternatum*		X	X	4-9
Broad Leaf Mountain Mint	*Pycnanthemum muticum*	X	X		3-9
Zig-Zag Goldenrod	*Solidago flexicaulis*	X	X	X	3-8

Source: http://www.nativeplanttrust.org /

Table 5: Primary Buffer Strip (Native Grasses)

Common Name	Scientific name	☀	⛅	☁	USDA Planting Zone
Big Bluestem	*Andropogon gerardii*	X			4-10
Sea Oats	*Chasmanthium latifolium*	X			5-8
Tufted Hairgrass	*Deschampsia cespitosa*	X	X	X	2-9
Wavy Hairgrass	*Deschampsia flexuosa*		X	X	4-7
Purple Love Grass	*Eragrostis spectabilis*	X	X		4-10
Switch Grass	*Panicum virgatum*	X			3-9
Little Bluestem	*Schizachyrium scoparium*	X	X		3-9
Indian Grass	*Sorghastrum nutans*	X			3-9

Source: http://www.nativeplanttrust.org /

Table 6: Secondary Buffer Strip (Native Trees)

Common Name	Scientific name	☀	⛅	☁	Gallons of stormwater uptake per year (Ohio) (4' Tree Diameter)	USDA Planting Zone
Large Trees						
Quaking Aspen	*Populus tremuloides*	X	X		146 Gallons	2-6
Eastern White Pine	*Pinus strobus*	X	X		185 Gallons	3-7
Eastern Hemlock	*Tsuga canadensis*	X	X		185 Gallons	3-7
Virginia Pine	*Pinus virginiana*	X			185 Gallons	4-8
Norway Spruce	*Picea abies*	X			185 Gallons	3-7
Tulip Tree	*Liriodendron tulipifera*	X	X		146 Gallons	5-9
Red Maple	*Acer rubrum*	X			116 Gallons	3-9
Northern Red Oak	*Quercus rubra*	X			143 Gallons	3-7
Black Walnut	*Juglandaceae*	X			146 Gallons	4-9
River Birch	*Betula nigra*	X	X		138 Gallons	4-9
Sweet Gum	*Liquidambar acalycina*	X			146 Gallons	6-9
American Sycamore	*Platanus occidentalis*	X			146 Gallons	5-9
Ohio Buckeye	*Aesculus glabra*	X	X		138 Gallons	3-7
Black Cherry	*Prunus serotina*	X			58 Gallons	3-9

Common Name	Scientific Name	☀	⛅	☁		USDA Zone
American Elm	*Ulmus americana*	X			82 Gallons	3-9
Understory Trees						
Fringe Tree	*Chionanthus virginicus*	X				4-9
Flowering Dogwood	*Cornus florida*	X	X		58 Gallons	5-9
Red Bud	*Cercis canadensis*	X	X		58 Gallons	4-9
American Holly	*Ilex opaca*	X	X		58 Gallons	5-9
Seaside Alder	*Alnus maritima*	X			69 Gallons	4-7
Serviceberry	*Rosaceae*	X			58 Gallons	4-8
Green Hawthorne	*Crataegus viridis*	X	X		171 Gallons	4-7
Flowering Almond	*Prunus triloba*	X	X			3-6

Source: https://mortonarb.org/ https://www.missouribotanicalgarden.org/PlantFinder/PlantFinderDetails.aspx?taxonid=281348 http://www.treebenefits.com/calculator/

Table 7: Secondary Buffer Zone (Native Shrubs)

Common Name	Scientific Name	☀	⛅	☁	USDA Planting Zone
Spicebush	*Lindera benzoin*	X	X	X	4-9
Smooth Blackhaw	*Viburnum prunifolium*	X	X		4-9
Witch Hazel	*Hamamelis virginiana*	X	X	X	4-8
Summer Sweet Bush	*Clethra alnifolia*	X	X		4-9
Bottlebrush Buckeye	*Aesculus parviflora*	X	X		4-8
American Cranberry Bush	*Viburnum opulus var. americanum*	X	X		2-6
Virginia Sweetspire	*Itea virginica 'Henry's Garnet'*	X	X		6-9
Mountain Laurel	*Kalmia latifolia*	X	X	X	5-9
Black Chokeberry	*Aronia melanocarpa*	X	X		3-8
Common Ninebark	*Physocarpus opulifolius*	X	X		3-7
New Jersey Tea	*Ceanothus americanus*	X			4-9
Shrubby Cinquefoil	*Potentilla fruticosa*	X	X		1-6
Shrubby St. John's wort	*Hypericum prolificum*	X	X		4-8
American Black Elderberry	*Sambucus canadensis*		X		4-9
Smooth Sumac	*Rhus glabra*	X	X		3-9
Black Huckleberry	*Gaylussacia baccata*	X	X	X	4-9
Lowbush Blueberry	*Vaccinium angustifolium*	X	X		3-8
American Hazelnut	*Corylus americana*	X	X		4-8

Source: http://www.nativeplanttrust.org /

Emergent plants are nature's biological filters and can naturally purify water better than high-tech artificial filters. The faster a plant grows and multiplies by rhizomes or seeds, the faster it purifies water.

Figure 7 & 8: Water lilies and cattails can filter water and help prevent eutrophication by filtering harmful stormwater nutrients.

Citation

"Ponds and Buffer Strips: Managing Stormwater Pollution." Ponds and Buffer Strips: Managing Stormwater Pollution, https://www.cooncreekwd.org/vertical/sites/%7B5C6B0F6F-9658-418B-9297-E0413AF79517%7D/uploads/%7BB32BFE59-E866-4D98-8B97-24DAF3D9CC45%7D.PDF. Accessed March 12 2022.

Table 8: Emergent Buffer Zone (Nutrient-Filtering Aquatic Plants)

Common Name	Scientific Name	☀	⛅	🔁	USDA Planting Zone
Green Arrow Arum	*Peltandra virginica*	X		X	5-9
Broadleaf Arrowhead	*Sagittaria latifolia*	X		X	4-11
Water Arum	*Calla palustris*		X	X	2-6
Blazing Star	*Liatris spicata*	X		X	3-8
Blue Cardinal Flower	*Lobelia siphilitica*			X	3-8
Blue Vervain	*Verbena hastata*			X	4-8
American Boneset	*Eupatorium perfoliatum*	X		X	4-5
Broadleaf Cattail	*Typha latifolia*	X		X	3-10
Common Rush	*Juncus effusus*	X		X	4-9
Greater Bur Reed	*Sparganium eurycarpum*		X	X	3-8
Joe Pye Weed	*Eutrochium purpureum*	X	X	X	4-9
Marsh Marigold	*Caltha palustris*	X	X	X	3-7
Swamp Milkweed	*Asclepias incarnata*	X		X	3-6
Smooth Phlox	*Phlox glaberrima*	X	X	X	3-8
Bush Monkey Flower	*Diplacus aurantiacus*	X	X	X	7-11

Obedient Plant	*Physostegia virginiana*	X		X	3-9
Pickerel Weed	*Pontederia cordata*	X		X	3-10
Porcupine Sedge	*Carex hystericina*		X	X	3-8
Water Smartweed	*Persicaria amphibia*	X	X	X	3-10
Water Iris	*Iris laevigata*	X		X	5-9
Fragrant White Water lily	*Nymphaea odorata*	X		X	4-10

Source: https://www.missouribotanicalgarden.org/

Citation

Withnall, Emily. Cattail: Plant Of A Thousand Uses, April 2, 2018.

Whetstone, J. w. (n.d.). AQUATIC & SHORELINE PLANT SELECTION.

Trees And Stormwater Calculator Tool — Urban Forestry South." Trees And Stormwater Calculator Tool — Urban Forestry South, Urbanforestrysouth.org, https://urbanforestrysouth.org/resources/links/trees-and-stormwater-calculator-tool.

"Hydrophytic Vegetation | Department Of Environmental Conservation." Hydrophytic Vegetation | Department Of Environmental Conservation, Dec.vermont.gov, https://dec.vermont.gov/watershed/wetlands/what/id/hydrophytes#:~:text=Obligate%20wetland%20plants%20include%20duckweed,Christmas%20fern%2C%20and%20Ground%20ivy..

USDA Plants Database." USDA Plants Database, Plants.sc.egov.usda.gov, https://plants.sc.egov.usda.gov/home.

Shaw, Danial. Plants for Stormwater Design, July 2003.

"Ponds and Buffer Strips: Managing Stormwater Pollution." Ponds and Buffer Strips: Managing Stormwater Pollution, https://www.cooncreekwd.org/vertical/sites/%7B5C6B0F6F-9658-418B-9297-E0413AF79517%7D/uploads/%7BB32BFE59-E866-4D98-8B97-24DAF3D9CC45%7D.PDF. Accessed March 12, 2022.

2.8: Rain Garden Design

A rain garden is a depressed area in the landscape that collects rainwater from a roof, parking lot, athletic field, tennis court, or roadway. They can also be located at the bottom of a hill or hills. Rain gardens allow water to soak into the ground. They are planted with the ideas of buffering the riparian zones with aquatic filtering plants, grasses, groundcover, native trees, and shrubs. Rain gardens can be a cost-effective and beautiful way to reduce stormwater runoff. Rain gardens can also help filter out pollutants in runoff and provide food and shelter for butterflies, songbirds, and other wildlife.

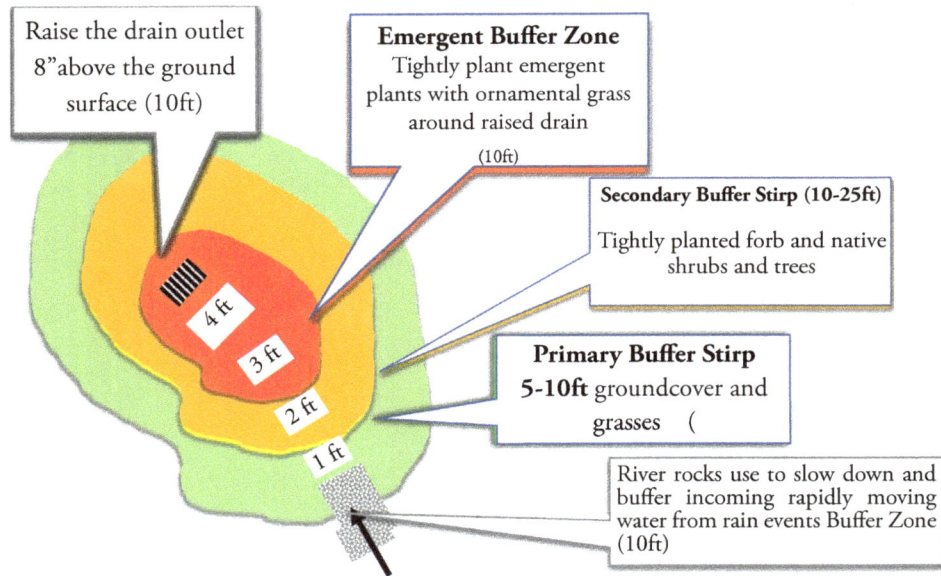

Figure 9: This is a diagram of how to install a rain garden with the same type of painting format as used in buffer strips.

Citations

Soak Up the Rain: Rain Gardens | US EPA." US EPA, Www.epa.gov, August 19. 2015, https://www.epa.gov/soakuptherain/soak-rain-rain-gardens.

Clark, Roberta. "Rain Gardens: A Way to Improve Water Quality." vol. 1, no. 26, Aug. 2011, ag.umass.edu/landscape/fact-sheets/rain-gardens-way-to-improve-water-quality. Accessed Aug. 2011.

2.9: Bioretention Design

Bioretention basins are landscaped depressions or shallow basins that slow and treat on-site stormwater runoff. Stormwater is directed to the basin and then percolates through the system, where several physical, chemical, and biological processes treat it. The slowed, cleaned water is then allowed to infiltrate native soils or directed to nearby stormwater drains or receiving waters. These systems are typically composed of seven elements that each carry out a different function.

Citation

Bioswales - USDA. 2005, https://www.nrcs.usda.gov/Internet/FSE_DOCUMENTS/nrcs144p2_029251.pdf.

2.10: Rainwater Collection

Cisterns have a greater storage capacity than rain barrels and may be located above or below ground. Due to their size and storage capacity, these systems (often large polyethylene drums) typically collect runoff from areas more extensive than residential rooftops, such as commercial parking lots. The collected water is generally used to irrigate landscapes, gardens, and ballparks regularly (i.e., feeding an automated irrigation system), reducing the strain on municipal water supplies during peak summer months. Cisterns may be used in series, and water is typically delivered using a pump system. Pump systems in cisterns can be designed with a floating level that shuts off the pump and converts the water source to a municipal supply when cistern levels are too low. Rain barrels are typically small-scale (25 to 100 gallons) and located at the downspout of a gutter system. They can also be linked to expand the overall storage volume. They are used to collect and store rainwater for watering landscapes and gardens or washing patio furniture. The simplest method of delivering water is by the force of gravity. However, more complex systems can be designed to provide the water from multiple barrels connected in a series with pumps and flow-control devices. The total storage volume available from rain barrels represents roof runoff from relatively small rainfall events, typically substantially less than one inch of rain over the surface. This is a small volume for a single rain barrel, but cumulative effects of rain barrels installed across a watershed include volume reduction and water-quality treatment since, typically, the first half to one inch of runoff contains the dirtiest water. During wet weather, there will likely be little or no storage available because of prior filling and little demand for irrigation water. If the ground can absorb it, consider discharging collected water onto vegetated areas between rainfall events to maximize rainwater capture and infiltration, even if unnecessary for irrigation.

Citation

Young, Edward S, and William E Sharpe. Rainwater Cisterns: Design, Construction, and Treatment, June 2, 2016, https://doi.org/https://extension.psu.edu/rainwater-cisterns-design-construction-and-treatment.

2.11: Recycling Stormwater and Nutrient Check

When irrigating from a stormwater retention pond before a rain event, you are reducing your stormwater nutrients by having your turfgrass absorb all the nutrients that have found their way into your pond, and the rainwater will replace the pond with fresh water. Think about it as a water change for a fish aquarium. To determine how many gallons a reservoir can hold, multiply length x width x average water depth (feet) x 7.5. At a minimum, efficiently irrigate 10 to 15 percent of the total volume biweekly. Test your water twice a month for pH, ammonia, nitrite, nitrate, and phosphorus. Note) It's important to keep the pH around 6.0-6.5 for freshwater ponds. Please see the safe testing numbers below.

1. Ammonia: 0.00 ppm - 0.10ppm
2. Total Nitrites/Nitrates: 0.00 ppm - 0.10ppm
3. Nitrates: 0.00ppm - 0.01ppm
4. Phosphorus: 0.00ppm - 0.25ppm
5. pH   6.0 - 7.5

Citation

Jackson, Sara. Keeping Up with Nitrate, 2018.

Rupard, Whitney D. "Testing Stormwater Chemistry A Watersafe Science Project ." Existing Stormwater Chemistry A Watersafe Science Project, 2015, https://doi.org/https://www.potomacriver.org/wp-content/uploads/2015/09/Testing-Stormwater-Chemistry-Student.pdf.

Archive Water Research Center - Phosphates. https://www.archive-water-research.net/index.php/phosphates.

PalaciosAn. Aquarium Water Change, How Often, How Much and Why, February 8, 2019.

Here are four pictures of recycling stormwater from a retention pond to water with a gas water pump (Figure 10) to irrigate a green space with an irrigated quick coupling system (Figure 11). Steve Gregory (Grounds Superintendent St Mary's College of Maryland) and a senior student in 2007 performed a phosphorus test from a lab spectrophotometer to ensure they were safe levels.

Figure 10

Figure 11

Figure 12

Figure 13

Figure 7 & 8: Here's a picture of the grounds team and a student testing the phosphorus levels from a stormwater retention pond we use to draw water to irrigate the turfgrass

2.12: Grounds Maintenance Stormwater Tips

Here are some tips for managing stormwater.

- Have a soil test done to understand soil nutrient needs.

- Use slow-release water insoluble nitrogen (WIN) or what about organic solids or teas?

- Use disease-resistant turfgrass cultivars to minimize pesticide spraying.

- Aerate at a six-inch depth and top-dress with USGA sand or spec compost for compacted soils to help lawns be more permeable.

- Clean drains consistently to remove silt and debris.

- Plant large canopy trees in parking lot islands to cool asphalt or concrete surfaces.

2.13: Environmental Impact and Cost Savings

Putting just a few checks and balances in place ensures that grounds-maintenance decisions are helping to preserve many aquatic and amphibian species by reducing algae blooms in the local watershed. Perhaps these efforts will increase tourism, fishing catches, and the quality and availability of seafood. It will also contribute to restoring the local water basin to healthy standards. By adding buffer zones around large water features within campus properties, maintenance needs, and fuel and labor costs will be reduced by not mowing and trimming spaces with a buffer area installed.

Table 9: Stormwater Prevention Checklist

Item	Yes	No	N/A
BMP program in place			
IPM program in place			
Buffer strips in place for all streams/ponds/rivers on the property			
Spill kits in place for fuel pumps/hazardous materials storage			
Secondary containment in place for pesticide			
Secondary containment in place for oil			
Secondary containment in place for fuel			
Plans in place to lower (WSN) synthetic fertilizer and replace it with (WIN) higher rates			
Rain gardens/Bioretention areas in place from a topographic map			
Water testing conducted for phosphorus levels			
Goals in place to change to lower toxicity pesticides			
Stormwater is recycled through irrigation.			
Is organic fertilizer used more than 50% of the time?			
Are parking areas more than 50% permeable pavement?			
Are sidewalks more than 50% permeable pavement?			
Are lawn areas more than 90% turfgrass coverage?			
Are each buffer strip more than 50% native plantings?			
Score			

Stormwater Report Card

Good..14–16

Fair..10–13

Poor..1–9

Section 3: Biodiversity Habitats

Biodiversity habitats are the intricate ecosystems around the world that sustain a wide variety of plant and animal life. They come in all shapes and sizes, from Kapok Tree (Ceiba pentandra) towering over rainforests to the hot dry deserts. Each one of these landscapes play a vital role in maintaining the health of our planet. There are three things needed to develop a functional wildlife habitat, which is genetic diversity, species diversity, and ecological diversity.

Genetic Diversity refers to the diversity (or genetic flexibility) within species. Each individual species of wildlife possesses genes that are the source of its own unique features. In human beings, for example, the huge variety in people's faces or colors of eyes, shape of nose, and so on reflect each person's genetic individuality. The term genetic diversity also covers distinct populations of a single species, such as the thousands of breeds of dogs that descend from the grey wolf.

Species Diversity is a measurement of an ecosystem's species abundance in a particular area, as well as its stability to thrive. If an ecosystem has poor species diversity, it may not function properly or be proficient. A diverse species grouping also contributes to ecosystem diversity because native wildlife needs both habitat and food sources for each individual species. For example, the black-chinned hummingbird prefers a southwest climate, which is a semi-arid country with river groves and suburbs. These birds breed in many kinds of semi-open habitats in the lowlands, including beside streams and in towns, brushy areas, oak groves, and canyons. They feed mostly on nectar and insects to gain energy and protein. The southwest offers many types of native plants with flowering foliage and insects year-round, which makes this an ideal habitat.

Ecological Diversity: is an understanding of all ecosystems within a geographical location (i.e., deserts, forests, rainforests, coral reefs, rivers, etc.). Ecological diversity refers to the variation in the ecosystems found in a region or the variation in ecosystems across the planet. Understanding what does not belong in a natural ecological area, like invasive plants, is just as important for maintaining a healthy habitat for shelter, breeding, and food sources for native wildlife.

Developing an ecological area within a landscape can range from something as small as adding a butterfly or pollinator garden to adding several acres of native flowers, shrubs, and trees. Having a good understanding of how each ecological area works and functions is key to any landscape. Accepting responsibility for protecting native wildlife is essential for the health of a property as well as its native wildlife. Table 10 is a list of some of the endangered species from the United States Fish and Wildlife Services.

Citation

"What Is Biodiversity? Its Definition, Protection, Loss And CSR Commitments." Youmatter, Youmatter. world, May 14. 2020, https://youmatter.world/en/definition/definitions-biodiversity-what-is-it-definition-protection-loss-and-csr-commitments/.

Lakna. Difference Between Genetic Diversity and Species Diversity, October 30, 2017, https://doi.org/htthttps://pediaa.com/difference-between-genetic-diversity-and-species-diversity/ps://pediaa.com/difference-between-genetic-diversity-and-species-diversity/.

Table 10: Some Examples of Endangered Species and Their Habitats

Endangered Species	Scientific Name	Locations	Habitat
Bumble bee, Rusty patched	*Bombus affinis*	Connecticut, Delaware, Washington, DC, Georgia, Illinois, Iowa, Kentucky, Maine, Maryland, Massachusetts, Michigan, Minnesota, Missouri, New Hampshire, New York, North Carolina, North Dakota, Ohio, Pennsylvania, Rhode Island, South Carolina, South Dakota, Tennessee, Vermont, Virginia, West Virginia, Wisconsin	Dry, dark cavities in wood
Monarch butterfly	*Danaus plexippus*	All U.S. states	Trees, shrubs, or other sheltered areas
Butterfly, Bay checkerspot	*Euphydryas editha bayensis*	California	Shallow, developed soil
Butterfly, Karner blue	*Lycaeides Melissa samuelis*	Illinois, Indiana, Michigan, Minnesota, New Hampshire, New York, Ohio	Grassy leaves
Butterfly, Uncompahgre fritillary	*Boloria microcinema*	Colorado	Grassy leaves and crevices in rocks
Bog turtle	*Clemmys muhlenbergii*	Connecticut, Delaware, Georgia, Maryland, Massachusetts, New Jersey, New York, North Carolina, Pennsylvania, South Carolina, Tennessee, Virginia	Edges of woods, wetlands, meadows

Turtle, Plymouth Redbelly	*Pseudemys rubriventris bangsi*	Connecticut, Maryland, Massachusetts, New Hampshire, New York, North Carolina, Pennsylvania, Rhode Island, Virginia, West Virginia	Wetlands, streams, ponds
Crane, whooping	*Grus americana*	Alabama, Arkansas, Colorado, Florida, Georgia, Idaho, Illinois, Indiana, Iowa, Kentucky, Louisiana, Michigan, Minnesota, Mississippi, Montana, North Carolina, New Mexico, Ohio, South Carolina, Tennessee, Utah, Virginia, Wisconsin, West Virginia, Wyoming	Shallow water
Woodpecker, red-cockaded	*Picoides borealis*	Alabama, Arkansas, Florida, Georgia, Louisiana, Michigan, North Carolina, Oklahoma, South Carolina, Texas, Virginia	Cavities in large, old pine trees

Source: https://www.fws.gov/endangered/

3.1: Types of Habitat Areas

When habitat areas are incorporated within a landscape, biodiversity wildlife for endangered species is promoted. In ecology, the term habitat summarizes the array of resources that promote the existence and propagation of a particular species. A habitat can be seen as the natural appearance of its environmental function. There are two factors needed to make a habitat area functional in a cultural landscape: environmental conditions and biotic resources. The environment is soil, moisture, range of temperature, and light intensity, and biotic resources are the availability of food and the existence or absence of predators.

The following sections contain ideas on how to incorporate ecological areas on campus.

3.2: Butterfly Garden Ecological Areas

There are more than twenty butterflies and moths on the endangered species list maintained by the U.S. Fish and Wildlife Services. Creating a butterfly garden can be accomplished in even a small setting. There are two types of plants needed for a butterfly garden to function well. Nectar plants are the plants that butterflies like to feed on, and host plants are plants that butterflies lay their eggs on and the caterpillars like to eat. Whether the butterfly garden needs long, grassy, native shrubs or small, loose stacks of rocks with crevices will depend on the type of butterflies that are desired. Different butterfly species prefer different nectar plants, and some like more than one. Providing several different types of nectar plants that have varying blooming stages is the ideal way to attract butterflies throughout the season. Be sure to have several types that bloom in the late summer and early fall because that is when butterflies are the most abundant. Fall-blooming natives such as asters and goldenrods are very important. Monarchs need as much fuel as they can get to power their long flight back to Mexico in the fall. In addition, check wind conditions and other weather parameters before pesticide application.

Some plants that are good for butterfly gardens and how they are propagated include:

- Milkweed (*Asclepias spp.*) milkweed seeds
- Butterfly weed (*Asclepias tuberosa*) butterfly weed seeds
- Butterfly bush (*Buddleia davidii*)
- Joe Pye weed (*Eupatorium spp.*), found in unique perennials
- Phlox (*Phlox spp.*)
- Ironweed (*Vernonia spp.*)
- Zinnia flowers
- Mexican sunflowers
- Agastache 'Ava'
- Brazilian verbena
- Callistemon spp. (bottlebrush)

Citations

Society, The Xerces. Gardening for Butterflies, 2016. "Butterfly Gardens." Butterfly Gardens - Gardening Solutions - University of Florida, Institute of Food and Agricultural Sciences, https://gardeningsolutions.ifas.ufl.edu/design/types-of-gardens/butterfly-gardens.html.

Figure 14: A Common Buckeye butterfly (Junonia coenia) pollinating a Sulphur Cosmos wildflower.

Figure 15: Western Honey Bee Apis mellifera pollinating Mammoth Grey Stripe Sunflower

Figure 16: Western Honey Bee pollinating a Sulphur Cosmos wildflower

3.3: Pollinator Ecological Areas

Pollination is a key part of a healthy environment. Plants provide food and shelter for many animals, and pollination helps plants thrive. This creates a balanced ecosystem that supports all sorts of life. Without pollination, these plants wouldn't be able to reproduce and make the seeds and fruits we harvest. Colony Collapse Disorder (CCD) is a complex phenomenon affecting honeybee colonies worldwide, characterized by the sudden disappearance of adult worker bees from hives. This leaves the queen, immature bees, and food stores behind, but the colony cannot function and eventually dies.

The exact causes of CCD are still not fully understood, but it is a combination of multiple factors, including:

Habitat loss: The loss of flowering plants and natural areas due to agricultural development, urbanization, and deforestation reduces the availability of food sources for bees.

Pesticides: Exposure to pesticides used in agriculture can harm bees directly or indirectly by reducing their food sources and weakening their immune systems.

Diseases and parasites: Varroa mites are a major parasite of honey bees that can weaken them and transmit diseases. Other diseases like bee viruses and fungi can also contribute to colony decline.

Climate change: Extreme weather events, such as droughts and floods, can disrupt bee foraging and colony survival.

When bees, insects, and mammals pollinate, they look for two things within the plant: nectar and pollen. Pollen is a protein, and nectar is a carbohydrate. For the most part, pollen is fed to bee larvae and queen bees. Nectar is the regular daily food for workers and drone bees. It is stored in the honeycomb, dehydrated, and made into honey. Butterflies, hummingbirds, bats, and other pollinators use nectar for energy to fly. When pollinators visit flowering plants, the pollen (which contains sperm) comes from the male plant, and the female structures have one or more ovaries (which contain eggs known as ovules). The pollen is carried from the pollinator on its wings, legs, or body to the female plant to reproduce.

In the United States, there are more than 4,000 species of native bees. Familiar bees visiting garden flowers are the colorful, fuzzy, yellow-and-black striped bumblebees, metallic-green sweat bees, squash bees, and imported honeybees. These flower-seeking pollen magnets purposefully visit flowers to collect pollen and nectar for food for themselves and their young ones. The European honeybee (Apis mellifera) is known for its importance in honey production.

There are two groups of bees that distinguish themselves from each other. The first group is called a tribe, or social bee, and the second group is called a solitary bee. Tribes such as bumblebees, honeybees, and stingless bees are from the super-family Apoidea and live together in colonies. Below is a description of how super-family bees function as a unit.

- Queen: The queen is the female at the center of the hive. The queen gives birth to the larvae that become the rest of the hive. The larvae are tended to by other members of the hive and stay in one spot.

- Worker: Worker bees are usually females, too, but these bees tend to be the queen, providing food, taking the eggs, caring for the larvae, and protecting the progeny.

- Drone: Drones are generally males and fly out in search of flowers to pollinate and obtain nectar to bring back to the colony for use with the workers, queen, and larvae.

Solitary bees usually have left the hive and are now alone. There are more than 200 species of solitary bees, and, as the name suggests, they live alone, although in truth, they often nest close to one another. They do not produce honey or have a queen and do not live in a beehive.

Typically, bees are not aggressive under the right conditions, but there are a few things that can irritate them.

Bee aggression could be caused by feeling threatened, raining, cloudy or windy days, starvation if there is lack of pollen and the nectar from lack of pollinator plants or no queen bee present.

Bad Weather: Bees prefer calm days with abundant sunshine. They do not care for cloudy, windy, or rainy weather. These types of conditions greatly affect the bees' temperament. Starvation: Hungry and thirsty bees will get irritated if there is not plentiful nectar and pollen available from flowers. Also if the queen bee goes missing through death or other misfortune, the hive will quickly know that she is gone. As they work to re-queen the hive, they can become more protective and defensive until the new queen appears.

There are other pollinators than just bees. Here are a few examples of some of them.

Beetles and Flies: Some species of flies are critical native pollinators, and many are mistaken for bees. Often, the coloring of flies will mimic those of bees, such as being striped in yellow and black.

Butterflies: Next to honeybees, butterflies are getting the most attention of the threatened insects, probably because they're so beautiful and beloved. They are incredibly important as native pollinators, too, and their hard work deserves notice. There are more than 700 species of butterflies in North America. The butterfly life cycle is closely tied to certain plant species that support the growing creature at various life stages. When habitats and plants are eliminated, so are butterflies.

Hummingbirds: These woodland birds are native pollinators and migratory, moving south for the winter and feeding along their migration routes. As they feed on nectar-producing plants, they pollinate them. Some species of plants have evolved to rely on the hummingbird for pollination.

Parasitic Wasps: These tiny insects are attracted to pollen and nectar, just like honeybees, and they are incredibly beneficial. Parasitic wasps consume aphids, scales, and flies, to name a few, helping restore balance in a garden. Gardeners who know them are grateful for their presence, allowing many home gardeners to go without chemicals and instead support wasp populations. They're called parasitic because their eggs are laid on insects such as the hornworm caterpillar, and upon hatching, the young wasps consume their host.

Source: https://www.treehugger.com/how-identify-different-types-bees-4864333

Citation

Pollinators In North America Committee, Status Of, et al. Status Of Pollinators In North America, 200 "Colony Collapse Disorder | US EPA." US EPA, Www.epa.gov, August 29. 2013, https://www.epa.gov/pollinator-protection/colony-collapse-disorder.

Donkersley, Philip. "Bees: How Important Are They And What Would Happen If They Went Extinct?." The Conversation, Theconversation.com, August 19. 2019, https://theconversation.com/bees-how-important-are-they-and-what-would-happen-if-they-went-extinct-121272#:~:text=Bees%20%E2%80%93%20including%20honey%20bees%2C%20bumble,vegetables%2C%20seeds%20and%20so%20on..

Motivan, E. (2020, February 14). What are some pollinating animals -- other than bees? ZME Science. Retrieved March 16, 2022, from https://www.zmescience.com/ecology/animals-ecology/incredible-pollinating-animals-bees/

Status of Pollinators in North America. National Academies Press, 2007.

Eleven explanations for aggressive honeybees (plus how to calm them!). Backyard Beekeeping 101. (2021, July 22). Retrieved March 16, 2022, from https://backyardbeekeeping101.com/aggressive-honey-bees/

Source: https://www.treehugger.com/how-identify-different-types-bees-4864333

"Habitats - Extension.umaine.edu." Understanding Native Bees, the Great Pollinators, University of Main, https://extension.umaine.edu/publications/wp-content/uploads/sites/52/2015/04/7153.pdf."How To Identify Different Types Of Bees." Treehugger, Www.treehugger.com, February 10. 2022, https://www.treehugger.com/how-identify-different-types-bees-4864333.

Wright, R., Mulder, P., & Reed, H. (2016). Honey Bees, Bumble Bees, Carpenter Bees and Sweat Bees, 1–10. https://doi.org/https://extension.okstate.edu/fact-sheets/honey-bees-bumble-bees-carpenter-bees-and-sweat-bees.html

Rau, Phil. The Biology and Behavior of Mining Bees, Anthophora abrupta and Entechnia taurea. Psyche 36(3):155-181, 1929.

Rourke, K. (n.d.). U.S. Forest Service. Forest Service Shield. Retrieved March 15, 2022, from https://www.fs.fed.us/wildflowers/pollinators/pollinator-of-the-month/anthophora-abrupta.shtml

U. (n.d.). https://leasehoney.com/2020/12/15/native-bees-of-america/. Retrieved from https://www.fs.fed.us/wildflowers/pollinators/pollinator-of-the-month/anthophora-abrupta.shtml

The native bees of America. LeaseHoney. (2020, December 17). Retrieved April 12, 2022, from https://leasehoney.com/2020/12/15/native-bees-of-america/

Motivan, E. (2020, February 14). What are some pollinating animals -- other than bees? ZME Science. Retrieved March 16, 2022, from https://www.zmescience.com/ecology/animals-ecology/incredible-pollinating-animals-bees/

3.4: Bee Identification

Identifying bees can be complicated because there are over 20,000 bee species in the world. There are some common characteristics that can help you narrow it down. All bees have three main body parts: head, thorax (middle section), and abdomen. They also have four wings, with the rear wings smaller than the front wings. Most bees are covered in hair, which helps them collect pollen. Below are some examples of endangered bees and some of the most common bees in the U.S.

Markings: The hair on the top of the head is black, short, and even. Males have a yellow abdomen with a black head and black striping in the lower thorax.

Habitat: Nest sites vary between bumblebee species. Most common species prefer dry, dark cavities, and nests can turn up in various ways from nests underground.

Such as abandoned rodent holes, under sheds, and compost heaps. Some make nests in thick grass, while others make nests in bird boxes, lofts, and trees.

Aggression: Bumble bees are typically very docile insects but can become defensive if necessary, so if you encounter a bumble bee or a nest, move slowly and with caution.

Figure 17: **American Bumble Bee** (*Bombus pensylvanicus)*

Figure 18: **The Bump Miner Bee** *(Andrena nasonii)*

Markings: Andrenas often look "furry," with particularly hairy thoraxes, faces and legs. The hairs covering the bees may be black, grayish, pale, brown or rust-colored. The bees' abdomens appear dark and are sometimes striped with pale bands.

Habitat: Miner bees prefer areas of bare soil with scattered vegetation, i.e., spotty growing grass areas, and preference for the tops of bare slopes. Their burrows are about 4-6 inches in length.

Aggression: Miner bees are docile bees that rarely sting unless threatened.

Figure 19: **Western Honeybee** *(Apis mellifera)*

Markings: Honeybees are around ½ inch long and have brownish bodies with bands of pale hairs on the abdomen.

Habitat: Honeybees live in beehives. They build their nest in gardens, woodlands, orchards, meadows, and other areas where flowering plants are abundant. Beehives can be found in honeybees inside tree cavities and under the edges of shelter objects to hide themselves from predators.

Aggression: Honeybees generally attack only to defend their colony or if they feel a threat to their nest from vibrations, smoke, etc.

Markings: They are dark brown to black, and many species have dark metallic green sheens. They have bands of hair on the outermost edge of their abdomen.

Habitat: They dig burrows underground in sandy soils and decaying wood.

Aggression: Furrow bees are generally not aggressive toward humans, except when they thirst for salt from humans. They will only sting if directly threatened.

Figure 20: **Furrow bees** *(Genus halictus)*

Markings: They have primarily black abdomen and are distinguished by visible yellow markings on the head.

Habitat: Moist to wet areas of deciduous woodlands, swamps, soggy thickets, wet prairies, marshes, seeps, and borders of streams.

Aggression: Fringed Loosestrife bees are docile bees and rarely sting unless threatened.

Figure 21: **Fringed Loosestrife Oil Bee** *(Macropis ciliata)*

Markings: Fine black hairs on the head, thorax, and under the abdomen.

Habitat: Grassy areas, woodlands, and soft dirt/sand.

Aggression: Males don't have stingers, but females do. They typically will not sting unless provoked.

Figure 22: **Long horned Bee** (Melissodes dentiventris)

Markings: Miner bees are smaller than honeybees with stout, furry bodies. They are frequently mistaken for bumble bees, also being black and yellow summertime bees.

Habitat: Miner bees are solitary, ground-nesting bees that like to establish their home in dry sand soils, grassy banks, and soils that are easy to dig. They can also be found in decaying wooden structures such as barns or cabins.

Aggression: Miner bees are docile and friendly and rarely sting unless threatened.

Aggression: Miner bees are friendly bees who usually will not sting unless provoked.

Figure 23: **Miner Bee** (*Anthophora abrupta)*

Markings: Males have a rust-colored "patch" on the upper middle part of the second abdominal segment, bordered by yellow along the sides and bottom. The U.S. Fish and Wildlife Service listed the rusty-patched bumble bee as endangered under the Endangered Species Act. If you spot one, please call your local government wildlife conservation office.

Habitat: The Rusty Patched prefers prairies, woodlands, marshes, agricultural landscapes, and residential parks and gardens

Aggression: Rusty-Patched bumble bee is non-aggressive and will only sting if provoked.

Figure 24: **Rusty Patched Bumble Bee** *(Bombus affinis)*

Markings: Golden Green Sweat bees are typically golden green but can range from a metallic blue to a tannish pink.

Habitat: Sweat bees nest in the ground or nest in rotten wood that is near grassy areas, gardens, and abundant flowering plants

Aggression: Sweat bees are non-aggressive and will only sting if provoked.

Figure 25:**Golden Green Sweat Bee** (*Augochlorella aurata*)

Source: https://www.treehugger.com/how-identify-different-types-bees-4864333

Citation

"Habitats - Extension.umaine.edu." Understanding Native Bees, the Great Pollinators, University of Main, https://extension.umaine.edu/publications/wp-content/uploads/sites/52/2015/04/7153.pdf.

"How To Identify Different Types Of Bees." Treehugger, Www.treehugger.com, February 10. 2022, https://www.treehugger.com/how-identify-different-types-bees-4864333.

Wright, R., Mulder, P., & Reed, H. (2016). Honey Bees, Bumble Bees, Carpenter Bees and Sweat Bees, 1–10. https://doi.org/https://extension.okstate.edu/fact-sheets/honey-bees-bumble-bees-carpenter-bees-and-sweat-bees.html

Rau, Phil. The Biology and Behavior of Mining Bees, Anthophora abrupta and Entechnia taurea. Psyche 36(3):155-181, 1929.

Rourke, K. (n.d.). U.S. Forest Service. Forest Service Shield. Retrieved March 15, 2022, from https://www.fs.fed.us/wildflowers/pollinators/pollinator-of-the-month/anthophora-abrupta.shtml

U. (n.d.). https://leasehoney.com/2020/12/15/native-bees-of-america/. Retrieved from https://www.fs.fed.us/wildflowers/pollinators/pollinator-of-the-month/anthophora-abrupta.shtml

The native bees of America. LeaseHoney. (2020, December 17). Retrieved April 12, 2022, from https://leasehoney.com/2020/12/15/native-bees-of-america/

Motivan, E. (2020, February 14). What are some pollinating animals -- other than bees? ZME Science. Retrieved March 16, 2022, from https://www.zmescience.com/ecology/animals-ecology/incredible-pollinating-animals-bees/

Figure 26: Non-Modified habitat with natural vegetation will provide nesting for wildlife.
Mow only once a year in late fall.

3.5: Non-Modified Ecological Areas

A non-modified naturalized area is undisturbed native vegetation only mowed once a year. These areas can be made up of grasses, weeds, native shrubs, and trees. The biodiversity from a non-modified naturalized area can sustain healthy bird populations by providing a variety of insects, seeds, and fruit for them to eat. The large biomass of insects supported by the naturalized area is essential for most native baby birds because they only eat insects, which provide the necessary protein for their rapid growth. Bees gathering nectar from native flowering plants is a good indicator of a healthy, balanced habitat. The presence of stoneflies and mayflies suggests clean, cold water with a high oxygen content along a freshwater shoreline, which is especially important from a stormwater management BMP perspective.

Naturalizing Your Local Park or Backyard - Lrconline.com. http://lrconline.com/Extension_Notes_English/pdf/Naturalize.pdf.

3.6: Semi-Modified and Modified Egologic/Pollinator Areas

Modified pollinator areas are artificial, altered landscapes designed to support and increase pollinator populations. These areas are crucial because many natural habitats for pollinators have been lost due to urbanization and agriculture. When incorporating semi-modified or modified wildflowers, follow these guidelines: Plant different cultivars of wildflowers that bloom throughout the growing season. This will provide continuous food sources for pollinators. Use pesticides sparingly because of their harmful and catastrophic effects on the bee colony. Pollinators need access to clean water for drinking and hydration. Installing wildflowers near streams, ponds, and stormwater retention is ideal.

Figure 27: Semi-modified wildflowers, which will include a mix of grasses, weeds, and perennial and annual wildflowers. These areas can offer color and function as pollinator areas for a fraction of the price.

49

Figure 28: Modified wildflower areas bloom throughout the growing season, and many have specific periods. To add more color and increase pollination efforts, add annuals to your native perennial wildflower mix in your modified areas. Sow spring annual wildflower seeds in late fall, and for summer annuals, sow seeds in late spring after the threat of the last frost.

Figure 29: Parking lot islands are a great way to promote pollinator areas, especially if you don't have a lot of green space. Once established, they also provide excellent color and low maintenance from mulching, trimming, and mowing grass.

3.7: Installing Bird and Bat Housing

Installing bird and bat houses aids in reducing insects like mosquitoes, flies, gnats, and other bugs since birds such as purple martins, blackbirds, bluebirds, sparrows, crows, and wrens consume such insects. Birdhouses offer alternative nesting sites in naturalized and could help boost populations of desirable bird species. The significant difference is minimizing disruption to the naturalized area. Ensure that birdhouses are made from untreated wood and avoid brightly colored options. Lasty birdhouses could attract predators like raccoons or snakes that might prey on nesting eggs inside the bird houses. Using predator baffles around the entrance bird houses holes can help moderate this.

Citation

Attracting Wildlife To Your Garden | University Of Maryland Extension." Attracting Wildlife To Your Garden | University Of Maryland Extension, Extension.umd.edu, June 3. 2021, https://extension.umd.edu/resource/attracting-wildlife-your-garden.

Table 12: Key Chart for Wildlife and Native Plant Biodiversity

🐦	Food Source / Migratory and Non-Migratory Bird
🕊	Food Source / Pollinator / Hummingbirds
🐝	Pollinators
🦋	Butterflies / Monarch Food Source
🐰	Small Mammals / Food Source
🦆	Waterfowl Food Source
🏠	Host Plant / Shelter / Nesting

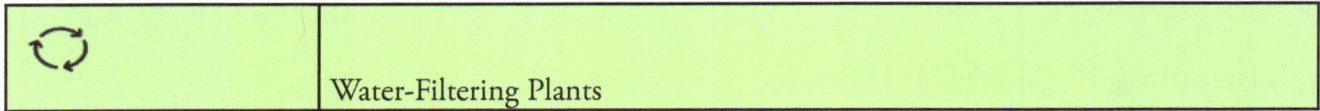

↻							
	Water-Filtering Plants						

Source:

Table 13: Primary Buffer Strip Zone (Native Groundcovers)

Common Name	Scientific Name	🐦	🐦	🦋	🐝	🏠
Maidenhair Fern	*Adiantum pedatum*			Yes		
Canada windflower	*Anemone canadensis*			Yes	Yes	
Red Bearberry	*Arctostaphylos uva-ursi*	Yes	Yes	Yes	Yes	
Goat's Beard	*Aruncus dioicus*			X	X	
Lady Fern	*Athyrium angustum*			X		
Bristle-Leaf Sedge	*Carex eburnea*		X	X	X	X
Pennsylvania Sedge	*Carex pensylvanica*			X		X
Blue Mist flower	*Conoclinium coelestinum*			X	X	
Tickseed	*Coreopsis rosea*			X	X	
Hayscented Fern	*Dennstaedtia punctilobula*			X		
Wild Strawberry	*Fragaria virginiana*	X	X	X	X	X
Wild Geranium	*Geranium maculatum*		X	X	X	X
Prairie Smoke	*Geum triflorum*			X	X	
Crested Iris	*Iris cristata*		X	X	X	X
Feathery False Solomon's Seal	*Maianthemum racemosum*	X	X	X		
Sensitive Fern	*Onoclea sensibilis*			X		
Cinnamon Fern	*Osmunda cinnamomea*			X		X
Royal Fern	*Osmunda regalis*			X		X
Running Groundsel	*Packera obovata*				X	X
Obedient Plant	*Physostegia virginiana*		X	X	X	
Woodland Stonecrop	*Sedum ternatum*	X	X	X	X	X
Broadleaf Mountain Mint	*Pycnanthemum muticum*			X	X	
Zig-Zag Goldenrod	*Solidago flexicaulis*	X	X	X	X	X

Source: http://www.nativeplanttrust.org/

Table 14: Primary Buffer Strip Zone (Native Grasses)

Common Name	Scientific Name	🐦	🐦	🦋	🐝	🏠
Big Bluestem	*Andropogon gerardii*	X				X
Sea Oats	*Chasmanthium latifolium*			X		X
Wavy Hair Grass	*Deschampsia flexuosa*			X		X
Purple Lovegrass	*Eragrostis spectabilis*	X		X		x
Vanilla Sweet Grass	*Hierochloe odorata*	X				X
Switchgrass	*Panicum virgatum*	X	X	X		X
Little Bluestem	*Schizachyrium scoparium*			X		X
Indian Grass	*Sorghastrum nutans*	X		X		X

Source: http://www.nativeplanttrust.org/

Table 15: Secondary Buffer Strip Zone (Trees)

Common Name	Scientific Name	🐦	🐦	🦋	🐝	🏠
Large Trees						
Balsam fir	*Abies balsamea*	X				X
Eastern White Pine	*Pinus strobus*	X				X
Eastern Hemlock	*Tsuga canadensis*	X				X
Jack Pine	*Pinus banksiana*	X				X
Norway Spruce	*Picea abies*	X				X
Willow Oak	*Quercus phellos*	X			X	X
Red Maple	*Acer rubrum*	X				X
Northern Red Oak	*Quercus rubra*	X				X
Wildfire Blackgum	*Nyssa sylvatica* 'Wildfire'	X				X
River Birch	*Betula nigra*	X				X
Yellow Popular	*Liriodendron tulipifera*	X	X	X	X	X
American Sycamore	*Platanus occidentalis*	X				X
American Beech	*Fagus grandifolia*	X				X
Black Cherry	*Prunus serotina*	X	X		X	X
American Elm	*Ulmus americana*	X				X
Understory Trees						
Fringe Tree	*Chionanthus virginicus*					

Flowering Dogwood	*Cornus florida*	X	X	X	X	
Red Bud	*Cercis canadensis*	X			X	
American Holly	*Ilex opaca*	X				X
Seaside Alder	*Alnus maritima*	X				X
Serviceberry	*Rosaceae*	X	X	X	X	X
Green Hawthorne	*Crataegus viridis*	X		X	X	X
Beach Plum	*Prunus maritima Marshall*	X				X
Striped Maple	*Acer pensylvanicum*	X				X

Source: https://www.feis-crs.org/feis/ https://mortonarb.org/

Table 16: Secondary Buffer Strip Zone (Native Shrubs)

Common Name	Scientific Name	🐦	🕊	🦋	🐝
Spicebush	*Lindera benzoin*	X	X	X	X
Smooth Blackhaw	*Viburnum prunifolium*	X	X	X	X
Witch Hazel	*Hamamelis virginiana*			X	X
Summer Sweet	*Clethra alnifolia*		X	X	X
Bottlebrush Buckeye	*Aesculus parviflora*		X	X	
American Cranberry Bush	*Viburnum opulus var. americanum*	X	X	X	X
Virginia Sweetspire	*Itea virginica 'Henry's Garnet'*		X	X	X
Mountain Laurel	*Kalmia latifolia*		X	X	X
Black Chokeberry	*Aronia melanocarpa*	X	X	X	X
New Jersey Tea	*Ceanothus americanus*		X	X	X
Shrubby Cinquefoil	*Potentilla fruticosa*		X	X	X
Shrubby St. John's Wort	*Hypericum prolificum*			X	X
Black Elderberry	*Sambucus canadensis*	X	X	X	X
Smooth Sumac	*Rhus glabra*	X	X	X	X
Black Huckleberry	*Gaylussacia baccata*	X	X	X	X
Lowbush Blueberry	*Vaccinium angustifolium*	X	X	X	X
American Hazelnut	*Corylus americana*	X	X	X	X

Source: http://www.nativeplanttrust.org/

Citation

"Plant Finder." Plant Finder, Www.missouribotanicalgarden.org, https://www.missouribotanicalgarden.org/plantfinder/plantfindersearch.aspx.

"Native Plants." Audubon, Www.audubon.org, https://www.audubon.org/native-plants.

Wild Flower Society (http://plantfinder.nativeplanttrust.org), New England. "Native Plant Trust." Native Plant Trust, Plantfinder.nativeplanttrust.org, https://plantfinder.nativeplanttrust.org/Plant-Search.

"The Morton Arboretum | To Plant And Protect Trees for a Greener, Healthier, And More Beautiful World." The Morton Arboretum, Mortonarb.org, 1 March. 2022, https://mortonarb.org/.

Borgmann, Kathi L, and Amanda D Rodewald. Native Landscaping for Birds, Bees, Butterflies, and Other Wildlife, November 13, 2013, pp. 1–4., https://doi.org/https://woodlandstewards.osu.edu/sites/woodlands/files/imce/0013.pdf.

3.8: Beneficial Insects

Beneficial insects are important for a healthy ecosystem. They help to keep pest insect populations in check, and they pollinate many of the plants that we rely on for food. They are also called predators' natural enemies, are another great biological control method to introduce in ecological areas. Biological control is the useful action of governing pests that damage plant life. There are basically three functions of beneficial insects which is pollinating, predator and parasitoids. Pollinating insects help plants grow by transferring pollen between flowers. Predator insects eat pest insects. Parasitoids insects lay their eggs inside pest insects or their eggs, which kills the pest once hatches. Please see the table below for all kinds of beneficial insects and what type of best they feed on.

Krischik, Dr. Vera, and Laurie Schneider. Guide to Integrated Pest Management (IPM), Jan. 2020, pp. 4–8. https://doi.org/https://ncipmhort.cfans.umn.edu/sites/ncipmhort.cfans.umn.edu/files/2020-05/guide%20to%20integrated%20pest%20management%202020may.pdf.

Raupp, Mike, et al. Predators, February 23, 2022, https://doi.org/https://extension.umd.edu/resource/predators#:~:text=The%20most%20common%20insect%20predators,(click%20on%20links%20below).

Ballew, Justin. NATURAL ENEMIES: PREDATORS AND PARASITOIDS, 23, 2 May 2019, https://hgic.clemson.edu/factsheet/natural-enemies-predators-and-parasitoids/.

Table 17: Predator Menu

Pest	Plants Pests Like	Predator
Aphids	sweet pea, dogwood, yucca flowering summer plants, milkweed, spirea, snowball viburnum, pines, juniper, spruce, elm, apple, crabapple, rose, Norway maple, beech, willow, and many summer annuals and perennials	Lacewing Ladybeetle Parasitic Wasps
Carpenter Worm	Oak, elm, maple, willow, cottonwood, black locust, box elder, sycamore, and ash trees	Parasitic Wasps
Elm Leaf Beetle	Elm trees	Parasitic Flies Parasitic Wasps
Lace Bug	Azaleas	Lacewing Ladybeetle Parasitic Wasps
Mealy Bugs	Many woody ornamental plants and some herbaceous perennials, coral bells, flax grasses, gardenia, hibiscus, cedar, and juniper	Lacewing Ladybeetle Parasitic Wasps
Scales	Many trees and shrubs	Lacewing Ladybeetle Parasitic Wasps Parasitic Mites
Spider Mites	arborvitae, azalea, spruce, rose, lantana, marigolds, impatiens, salvia, and viola	Lacewing Ladybeetle Parasitic Mites
Thrips	gladiolus, impatiens, figs, irises, roses	Lacewing Parasitic Wasps Parasitic Mites
Mosquitoes	Mosquitoes are pests and can spread diseases such as West Nile Virus, dengue, and malaria to human life forms.	Mosquitoes Fish Purple Martins Dragonflies Bats
Beetles, grasshoppers, crickets, and other pest insects	Bushes are the preferred habitat for newly hatched praying mantids.	Praying Mantids.
Aphids, caterpillars, scale insects, spider mites, and insect eggs	Tree groves, forests, and lakes	Assassin Bugs

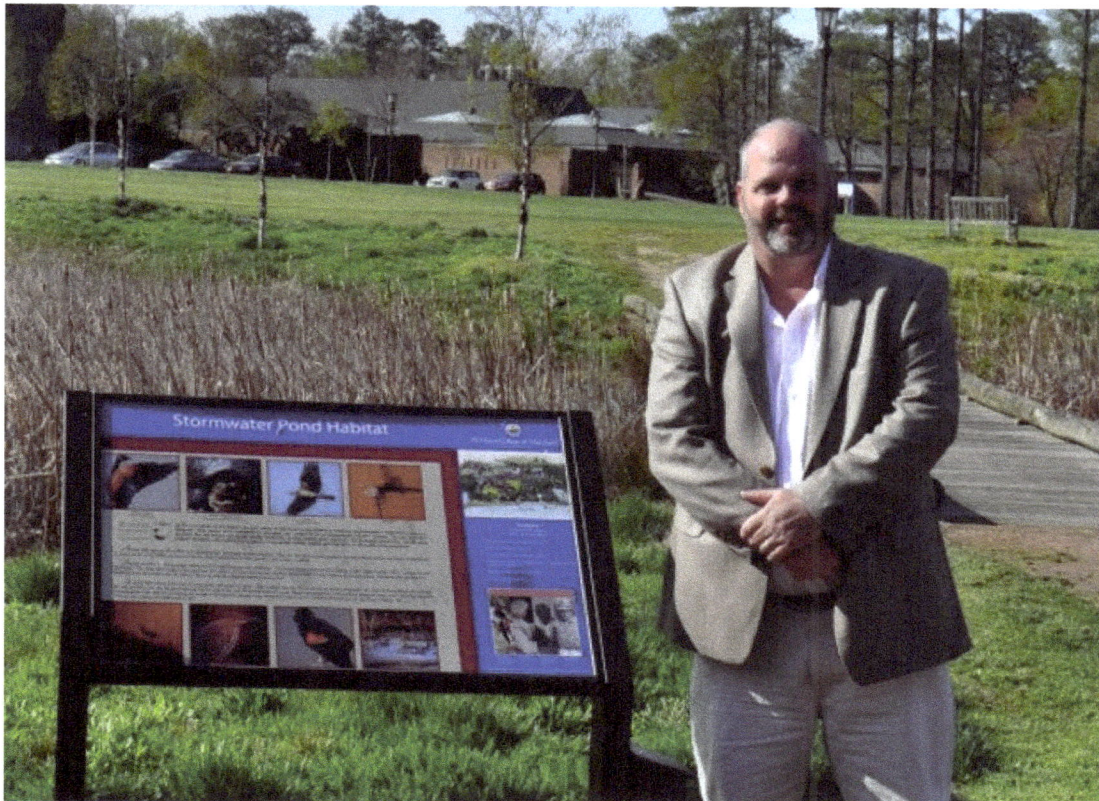

Figure 30: Encouraging public participation in wildlife observations through informative signage can lead to increased reporting of endangered species and beneficial pollinators.

3.9: Mechanical Weed Control

Mechanical weed control is a physical activity that impedes the growth of weed plants. The goal is to remove, injure, kill, or make the growing conditions for weeds or undesired plants unfavorable. There are pros and cons to each type of mechanical control method. Below are five types of mechanical methods.

1. **Hand pulling/Digging up weeds.**

   ○ **Pros:** Digging up or pulling weeds removes the entire weed, roots and all, from the ground. By taking out the whole plant, the same plant will not usually grow again in the same location. Individually removing weeds also ensures that existing plants are not damaged or accidentally killed, as compared to spraying a herbicide. The best way to dig up weeds is to wait until after it rains when the soil is still wet and soft enough to pull the weeds out.

   ○ **Cons:** Removing weeds individually by pulling or digging them from the ground is extremely labor-intensive, especially in large areas like planting beds.

2. Flaming

○ **Pros:** Using a flame torch for weed removal is good for damaging the plant down to its roots. The ashes blow away quickly, leaving no traces of dead weed present for days. Using a flame torch is less labor-intensive than pulling weeds.

○ **Cons:** Roots still exist underground, and weeds can come back and grow again. Typically, using a flame torch is more effective on broadleaf weeds compared to grassy weeds. Some local, state, and federal governments might require a permit. Safety is another issue. The most common injury to workers is accidentally burning their own feet.

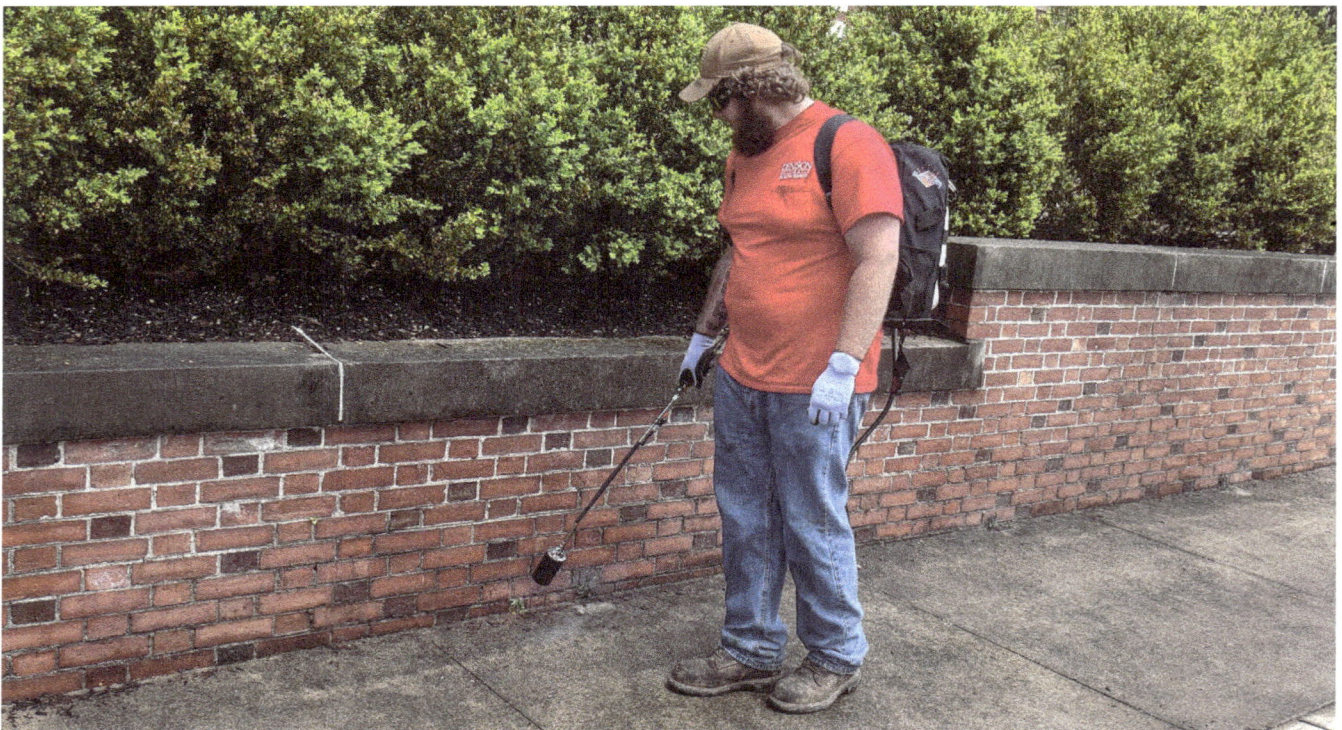

Figure 31: A Denison University employee using a propane blow torch to burn the grassy and broadleaf weeds on the campus sidewalks.

3. Pine Mulch

○ **Pros:** Pine straw is typically a great sustainable asset to apply to a landscape because only pine needles that fall from the trees are used. Several soil benefits occur when using pine needle straw mulch for beds and open areas of islands under large tree canopies or tree groves. The mulch conserves soil moisture because it reduces daily evaporation. It also insulates the roots of plants to prevent extreme changes in temperature. There are erosion-protection benefits when using pine needle straw mulch on medium- to steep-grade banks. It prevents soil compaction by reducing the impact of precipitation from the surface. When a thick layer of pine needle straw mulch is placed around plants, weeds struggle to survive. Wood mulch allows the seeds from weeds to germinate within the product, eventually creating the need to restore the garden through manual labor. The straw allows the weed seeds to fall through the product to a soil base that is just acidic enough to discourage growth. The pine needles tend to interlock with each other, holding fast to create a blanket of warmth that maintains adequate heat at the soil level. The added temperatures can also help some plants grow faster or stronger in the early season to produce better yields. When used with other mulch products, the manufacturers often recommend complete removal of the installed layer before replacing it, which allows the nutrients from the new mulch to reach the plants.

○ **Cons:** Pine straw offers very little nutrient value. Insects love nesting in pine straws due to the insulated cover they offer. Pine straw may present a fire hazard when drought occurs or when applied before the first rain event.

4. **Landscape Fabric**

○ **Pros:** Landscape fabric does a fairly good job of slowing down weed growth for a couple of years when installed correctly. Landscape fabrics work best in landscape beds that are designed to be permanent (i.e., shrubs, perennials, and small flowering trees). It also provides moisture and cooler temperatures for the soil during the hot summer months and warmer temperatures in the winter months. Landscape fabric does a great job of stabilizing soil for erosion control.

○ **Cons:** Landscape fabrics prevent nutrients from reaching the soil, which can be catastrophic for plants if the soil is not rich in nutrients. Another problem with landscape fabrics is weed seeds can end up growing in the mulch placed on top of the fabric, which defeats its objective. Also, some fabrics break down after a couple of years, so be careful when selecting landscape fabrics and remember the old saying, "Pay now or pay later."

5. **Goats**

○ **Pros:** Goats do a great job of controlling invasive species like Kudzu, Oriental bittersweet, Ailanthus, mile-a-minute, sumac, winged elm, and ironweed.

○ **Cons:** Goats cannot eat or consume poison hemlock, Johnsongrass, Sudangrass, milo, and sorghum-sudangrass. While goats do a fantastic job of eating invasive plants, they don't know the difference between invasive plants and perennials, shrubs, and trees. Goats also only eat the foliage of plants, so if plants have stolons and rhizomes, they will more than likely come back the following year.

Table 18: Biodiversity Checklist

Item	Yes	No	N/A
Are non-modified ecological areas in place?			
Are pollinator ecological areas in place?			
Are ecological buffer strips in place?			
Is there an inventory for non-migratory birds?			
Is there an inventory for migratory birds?			
Is there an inventory for monarch butterflies?			
Is there an inventory for predators vs. pest populations within the developed ecological areas?			
Is there an inventory for bees?			
Is there an inventory for hummingbirds?			
Is there an inventory for waterfowl?			
Is there a plan for invasive plant species?			
Is the local endangered wildlife list known?			
Is there signage for the ecological areas?			
Is there a pesticide buffer zone in place around all ecological areas?			
Are safeguards in place before applying any insecticides?			
Are non-neonicotinoid pesticides used?			
Are ecological areas part of the admissions tour?			
Are there managed pollinator protection plans (MP3) in place?			
Score			

Biodiversity Report Card

Good...14–16

Fair..10–13

Poor...1–9

Section 4: Carbon Fertility and Composting

Composting has been around for thousands of years. While the specific techniques have evolved over time, the basic concept of using organic matter to improve soil fertility has been a cornerstone of agriculture since its beginnings. Composting is the natural process of decomposing organic matter into nutrient-rich soil organic amendments, also known as compost. It's a great way to recycle food scraps, yard waste, and other organic materials that would otherwise end up in landfills. Compost could improve soil quality, add nutrients to plants, and help retain moisture. The benefits of composting can reduce waste and divert organic waste from landfills, where it decomposes anaerobically and produces methane, a potent greenhouse gas. Composting improves soil health by improving drainage, aeration, and water retention. It also feeds the beneficial microorganisms in the soil, which helps to break down nutrients and make them available to plants.

Soils perform at their peak for maximum crop production when their chemistry is balanced regarding macro and micronutrients. A completely balanced soil, with a carbon-based fertility program, activates nutrients more efficiently, provides better drainage by opening micropores, and aids in the checks and balances for pathogens and disease. Overall, carbon fertility saves money, labor, and headaches!

Adding carbon as organic matter provides readily available energy for soil microbes. These microbes help break down minerals into plant-usable forms, increasing the efficiency of nutrient utilization and reducing the need for synthetic fertilizers. Organic matter improves soil aggregation, creating air and water pockets and promoting healthy root growth and drainage. This reduces compaction and waterlogging, ultimately optimizing plant health and yield. A carbon-rich environment fosters a diverse microbial community, offering natural pests and disease control through predation and competition. This reduces reliance on chemical pesticides and promotes a more balanced ecosystem. By minimizing synthetic inputs and promoting natural nutrient cycling, carbon-based fertility programs contribute to lower greenhouse gas emissions and reduce water pollution.

Sunlight causes photosynthesis, the process of the sunlight energy captured by chlorophyll, which converts carbon dioxide and water into sugars and oxygen, producing food for the plants to live on. What the plant doesn't need for growth is exuded through the roots to feed soil organisms that make up healthy organic soils. Carbon is the main component of soil organic matter and helps give soil its water-retention capacity, its structure, and its fertility.

So, where does carbon-based fertility come into play, and what is it exactly? The available fraction of carbon (humus) in the soil is what makes the whole soil work, and it addresses one of the most basic agronomic principles: the carbon-to-nitrogen ratio which is typically has a ratio of 30 parts of carbon to 1 part nitrogen.

By building carbon through multiple application, you can balance and feed this important carbon-to-nitrogen ratio in the soil. This allows for a reduction of total nitrogen application, which helps the soil physically and biologically while also helping the grounds operational budget.

There are a number of commercial ways to get carbon into a lawn maintenance program but be warned that not all products are built the same. A good carbon-based fertilizer (organic fertilizer) goes through a process called digestion which aids in the microorganisms to flourish and look for the USDA certification label to know it's a true organic product.

Citation

Schwartz, Judith D. Soil as Carbon Storehouse: New Weapon in Climate Fight? March 4 2014, pp. 4–4., https://doi.org/https://e360.yale.edu/features/soil_as_carbon_storehouse_new_weapon_in_climate_fight.

Blum, B. (1992). Composting and the Roots of Sustainable Agriculture. Agricultural History, 66(2), 171-188.

Retrieved from https://www.jstor.org/stable/3743852? seq=1#metadata_info_tab_contents

Schader, Meg. "How Does the Sun Affect Plants?" September 30, 2021, pp. 1–4.

https://doi.org/

https://sciencing.com/what-happens-to-plants-if-they-have-no-sun-12486997.html

Callis, S. (2020). Another Short History of Composting. *Missouri State Coordinator*, 1–4.

There are six primary nutrients that plants require in fairly large quantities:

- Carbon from carbon dioxide in the air

- Hydrogen from water

- Oxygen from water and air

- Nitrogen helps plants make the proteins they need to produce new tissue. In nature, nitrogen is often in short supply, so plants have evolved to take up as much nitrogen as possible, even if it means not taking up other necessary elements. If too much nitrogen is available, the plant may grow abundant foliage but not produce fruit or flowers.

- Phosphorus stimulates root growth, helps the plant set buds and flowers, improves vitality, and increases seed size. It does this by helping transfer energy from one part of the plant to another. To absorb phosphorus, most plants require a soil pH of 6.5 to 6.8. Organic matter and the activity of soil organisms also increase the availability of phosphorus.

- Potassium improves the overall vigor of the plant. It helps the plants make carbohydrates and provides disease resistance. It also helps regulate metabolic activities.

There are also three additional nutrients that affect plants.

- Calcium is used by plants in cell membranes at their growing points and to neutralize toxic materials. In addition, calcium improves soil structure and helps bind organic and inorganic particles together.

- Magnesium is the only metallic component of chlorophyll. Without it, plants can't process sunlight.

- Sulfur is a component-specific amino acids, proteins, membranes, and coenzymes.

Citation

Zhao, Zhanhui, et al. "Effect of Compost and Inorganic Fertilizer on Organic Carbon and Activities of Carbon Cycle Enzymes in Aggregates of an Intensively Cultivated Vertisol." PLOS ONE, vol. 15, no. 3, 2020, https://doi.org/10.1371/journal.pone.0229644.

Kennedy, Taryn. "Compost Standers ." What Standards Must Compost Products Meet to Be Used in Organic Agriculture? May 2019, pp. 1–4. https://doi.org/https://www.omri.org/compost-standards.

"Turfgrass Fertilization." A Basic Guide for Professional Turfgrass Managers, November 10, 2016, pp. 3–4. https://doi.org/https://extension.psu.edu/turfgrass-fertilization-a-basic- guide-for-professional-turfgrass-managers.

Lal R.

Carbon sequestration.

Philos. Trans. R. Soc. Lond. B Biol. Sci. 2008; 363: 815-830

4.1: Soil Biology

Healthy soil is basically made up of nine individual types of organisms, bacteria, fungi, and beneficial nematodes. When repetitive pesticide and synthetic fertilizer applications are applied, they could disrupt and deplete these necessary biological components to promote optimum plant growth and health. Listed below are all the components that make up a well-balanced soil. Testing for these bacteria, microorganisms and fungi is expensive. However, it is very beneficial to see what things might need replenishing for your soil optimum plant health. To ensure your soil doesn't have any deficiencies a test must be completed from a specialized lab that can test for these microorganisms. True carbon-based products like properly made compost and certified carbon-based fertilizers products from the USDA could help restore your soil for optimum plant health.

- ○ **Psychrophiles:** These microorganisms thrive in cold environments, with an optimal growth temperature below 15 °C (59 °F). They play a role in decomposing organic matter in the soil. Some psychrophilic bacteria can return nitrogen from the atmosphere to the soil for vital nutrient for plants from soil uptake. They also help filter contaminated soils like pesticides, oil, etc.

- ○ **Mesophiles**: These microorganisms are the most common type found in the soil. These microorganisms decompose organic matter like plants and dead animals, which releases essential nutrients like nitrogen, phosphorus, and potassium back into the soil. These nutrients are then taken up by plants for their growth. The breakdown of organic matter will create humus, a dark, organic material that improves soil structure and helps make available nutrients for the plant. Some Mesophiles have mutual relationships with mycorrhizal fungi. These fungi help plants to absorb nutrients from the soil more efficiently. These microorganisms thrive in moderate temperatures, typically between 20°C and 45°C (68°F and 113°F) and will typically break down organic matter and generate heat. Mesophiles quickly break down readily available compost organic matter like food scraps and yard trimmings, releasing heat and initiating composting breakdown cycle. The heat will raise, and this is why its important to turnover the compost piles to avoid beneficial microorganisms being kill from the excessive heat.

- ○ **Thermophiles:** These microbes are appropriately named for their love of heat and are the dynamic microbes for composting. that thrive in high temperatures, typically between 41 and 122 °C (106 and 252 °F). These microorganisms can remain in hot temperatures found in compost piles which allows organic matter break down rapidly. This decomposition process releases nutrients back into the ecosystem, making them available for other organisms.

- • Hyphal Diameter/Fungi: Soil fungi are a dynamic part of healthy soil ecosystems. These microscopic, thread-like organisms aid in nutrient cycling, plant growth, and overall soil health. They help with breaking down organic matter, enhanced plant growth, stabilized soil structure and aid in suppressing disease pressure.

- ○ **Protozoa:** While bacteria and fungi often take the spotlight in discussions about composting, protozoa, tiny single-celled organisms, also play a significant role in creating healthy, nutrient-rich compost. This bacterium helps regulate bacterial population by feeding and by aiding from consuming wastewater. They help maintain fertile soils by releasing nutrients from what they eat. Protozoa also force very small fragments of aggregates within the soil as they search for bacteria to consume, which helps macropores and micropores.

- • Flagellates, Amoebae, and Ciliates (FAC): Among the diverse microscopic world of protozoa in compost, Flagellates, Amoebae, and Ciliates each play specific roles. They use whip-like flagella to propel themselves and capture bacteria, helping regulate bacterial populations and prevent excessive decomposition. Different flagellate species specialize in different food sources, contributing to a diverse and balanced microbial community. Protozoa also help build the larger soil pores by pushing aggregates around as the protozoa search for and try to reach the bacteria tucked away around soil particles.

- Beneficial Nematodes: These types of nematodes actively search for pest insects in the soil. Once they locate a suitable host, they enter the insect's body through natural openings or burrow through the insect's cuticle (outer shell). They help control a wide range of beetle grubs, caterpillars, cutworms, weevil larvae, cicada nymphs and mole crickets through.

<u>Citation</u>

Ingham, E. R. (n.d.). Soil Bacteria. THE LIVING SOIL: BACTERIA. https://doi.org/https://www.nrcs.usda.gov/wps/portal/nrcs/detailfull/soils/health/biology/?cid=nrcs142p2_053862

DeJong-Hughes, J. (n.d.). Soil Biology. Extension at the University of Minnesota. Retrieved March 16, 2022, from https://extension.umn.edu/soil-management-and-health/soil-biology#practices-that-increase-microbial-populations-139006

Lehmann, A., Zheng, W., Soutschek, K., Roy, J., Yurkov, A. M., & Rillig, M. C. (2019). Tradeoffs in hyphal traits determine mycelium architecture in saprobic fungi. Scientific Reports, 9(1). https://doi.org/10.1038/s41598-019-50565-7

Ingham, E. R. (n.d.). Soil Protozoa. THE LIVING SOIL: PROTOZOA, 1–2. https://doi.org/https://www.nrcs.usda.gov/wps/portal/nrcs/detailfull/soils/health/biology/?cid=nrcs142p2_053867#:~:text=Protozoa%20play%20an%20important%20role%20in%20mineralizing%20nutrients%2C%20making%20them,than%20the%20bacteria%20they%20eat.

Compost itself is decomposed organic matter. It can often be confused with fertilizer, but there is an important distinction between the two. Synthetic fertilizers, while they go into the soil, are intended to. Compost should be considered a soil amendment unless it is treated and manufactured in a pelletized form with manure, blood meal and bone meal added with the percentage of three to five percent nitrogen. Carbon fertility works by feeding microorganisms through the denitrification process within the soils.

4.2 Types of Composting

Compost is made up of carbon and nitrogen, creating a carbon-to-nitrogen ratio. It is important to know the carbon-to-nitrogen ratio of soil. Typically, a thirty-to-one carbon-to-nitrogen ratio is desirable. There are two types of inputs when making compost: green and brown. Greens represent nitrogen and are typically found in things like grass clippings, coffee grounds, tea bags, vegetable and fruit scraps, trimmings from perennial and annual plants, annual weeds that haven't set seed, eggshells, animal manure (cow, horse, sheep, chicken, rabbit, and seaweed/kelp). The browns represent carbon and are typically found in fall leaves, pine needles, sawdust, corn stalks, woodchips, etc.

To create a thirty-to-one carbon-to-nitrogen ratio, you will need 30 percent brown and 1 percent green material in your compost. Brown materials are carbon-based, and green materials are nitrogen-based (see Table 20). Approximately 25,000 square feet are needed to stockpile or windrow compost.

Citation

McKibben, W. L. (2021). A growers guide for balancing soils: A practical guide to interpreting soil tests. Acres U.S.A.

Brunetti, J. (2014). The farm as an ecosystem: Tapping nature's reservoir – biology, geology, diversity. Acres U.S.A.

Hoorman, J. J. (2010, September 7). Understanding soil microbes and nutrient recycling. Ohioline. Retrieved March 16, 2022, from https://ohioline.osu.edu/factsheet/SAG-16

Scalera, S., Reisinger, A. J., & Lusk, M. (2019). The importance of soil health for residential landscapes. EDIS, 2019(2). https://doi.org/10.32473/edis-ss664-2019

United States Environmental Protection Agency. n.d. "Composting at Home." Accessed September 24, 2021. https://www.epa.gov/recycle/composting-home.

Trautmann, N., & Olynciw, E. (n.d.). Compost Microorganisms, 1–5. https://doi.org/http://compost.css.cornell.edu/microorg.html

Hoidal, N. (2021). Should You Add Microbial Soil Amendments to Your Garden? https://doi.org/https://extension.umn.edu/yard-and-garden-news/should-you-add-microbial-soil-amendments-your-garden

The living soil - microorganisms. Vermont Organic Farm | Cedar Circle Farm & Education Center. (n.d.). Retrieved March 16, 2022, from https://cedarcirclefarm.org/tips/entry/the-living-soil-microorganisms

4.3: Composting

Stockpile Composting involves dumping nitrogen and carbon waste inputs into a pile and turning the entire pile over twice a month with a front-end loader. Turning is essential, as it helps to distribute organic materials uniformly. It also ensures heat buildup does not destroy the production of good fungi. After six months, stockpile compost should be broken down using a tumble screen machine, which will speed up the composting process and deliver a desirable thirty-to-one carbon-to-nitrogen ratio sooner.

Figure 32: Denison University uses compost to help promote healthy organics in the soil .Shown here is a screening machine that helps refine the compost to break down quicker when applying them on the campus lawns.

Aerated Windrow Composting also requires stockpiling; however, it begins with thoroughly mixing the compost pile with a front-end loader and then placing the compost into windrows around two to three feet high and fifty feet long. A windrow machine aerates the pile, improving porosity and oxygen content and allowing for moisture control and the redistribution of heat. The windrows are turned multiple times during the composting process, which takes an average of sixteen weeks, depending on maturity requirements. The steps below can be followed to create aerated windrow compost:

- Screen yard waste and animal manure and then mix with a front-end loader. Place the mixture into windrows along a non-permeable surface.

- Turn windrows on a regular basis to improve oxygen content and distribute heat to regulate temperature and moisture.

- Screen the compost again to remove contaminants and to grade for various end uses. Oversized materials can be removed and put back through the whole process again until they have been composted down sufficiently.

Figure 33: Here is a picture of a Denison University employee aerating compost using a compost turner machine.

Figure 34: Shown here is a Denison University groundskeeper applying compost to the campus lawns to help promote healthy soils.

4.4: Vermicomposting

Vermicomposting, composting with worms, is usually done with the common red wiggler worm (Eisenia fetida). This worm's specialized digestive system converts food waste and other organic materials to a nutrient-rich compost called vermicast, or worm castings. It thrives in an aerobic (with air) environment. It can process large amounts of food waste and rapidly reproduce in a confined space. Vermicomposting is a healthy and clean way to eliminate the waste going into landfills, which improves the environment. Vermicomposting is inexpensive and only takes two to three months to produce results. The typical nitrogen, phosphorus, and potassium analysis from vermicomposting averages around 3 percent nitrogen, 2 percent phosphorus, and 2 percent potassium. Vermicomposting reduces food waste by 30 percent. When vermicomposting, starting with a baseline mix is essential because organic products can be added to strengthen the mix depending on nutrition needs. A baseline mix includes beneficial enzymes. Some enzymes accelerate the breakdown of organic matter, such as hydrolase and glucosidase, while others promote nutrient mineralization, which breaks down the compounds in organic matter, like phosphates and sulfates, through an oxidation process. Once these compounds are broken down, they become soluble inorganic materials that can be made available to plants through micropores and macro availability.

When feeding the red wiggler worms used in vermicomposting, use brown leaves (70 percent) as a carbon source and any food waste (30 percent) as sources of nitrogen, phosphorus, and potassium. Remember, meat and dairy products are not recommended.

Citation

https://www.epa.gov/sustainable-management-food/types-composting-and-understanding-process#aeratedturned

The composting process. Composting in the Home Garden - Common Questions. (n.d.). Retrieved March 16, 2022, from https://web.extension.illinois.edu/compost/process.cfm

4.5: Compost Tea

Compost tea is a liquid fertilizer brewed from compost food waste. Compost tea is a way to introduce beneficial microbes and nutrients by adding beneficial microbes to the soil, improving plant growth, and suppressing diseases. Compost tea is relatively easy to make (please see steps one- six for instructions at home. There are many different recipes for N-P-K (please see Table 19 below).

Step 1

A worm composter is, at its simplest, a bin with holes for ventilation and moisture. It is almost always made with plastic and raised off the ground to allow water to drain out the bottom. Wigglers, white worms, and other earthworms create a mixture of decomposing vegetable or food waste, bedding materials, and vermicast. One pound of worms (about 1,000 worms) will eat about ½ to 1 pound of food scraps per day..

Step 2

An 800-gallon tank (shown here) works fine. Plum an air pump line using a 2" PVC pipe with several small drill holes into the pipes and installed it to an air pump.

Step 3

Fill up the tank with water. If tap or potable water is being used, bubble out the chlorine for at least 2-3 hours minimum. (Shown Here)

Step 4

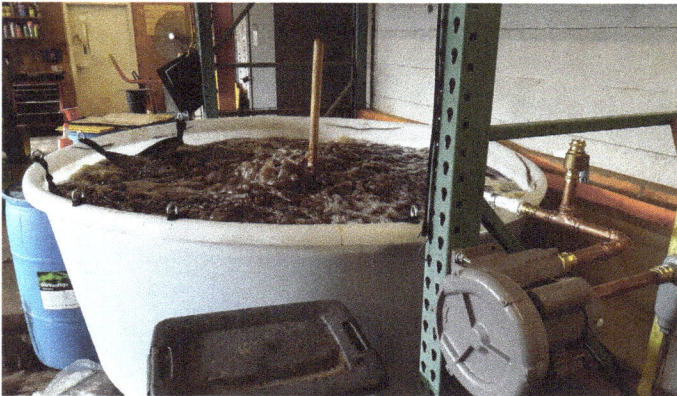

Once chlorine has been removed. Fill the tea bags with worms casting from your vermin casting bin (10lbs of worm casting per 100 gallons of water) Let it brew for 6 -8 hours.

Step 5

After the compost tea has brewed for 6-8hrs, add one gallon of molasses and any other organic or other sugar products so the enzymes can feed off them and multiply and let it brew for 24hrs. The foam brew head on top of the compost brew tea is an indicator that the enzymes are working correctly and efficiently.

Step 6

A transfer pumps works well to pump the compost tea into a spray rig. Bloomless spray nozzles can be used to drench areas at a rate of 80-100 gallon per acre. (Note) Be sure to rinse your spray rig out thoroughly to avoid killing enzymes and other biological matter.

Table 19: Food-Waste Sources for Compost Tea

Food Waste	N	P	K	Cal	Mag	Sulfur	Boron	Iron
Cucumbers, squash, zucchini, winter squash, mirliton, pumpkin, gourds, and cucuzzi	X	X	X		X		X	
Avocado, kiwi, papaya, tomato, chestnut, potato, and bell pepper	X				X		X	
Coffee grounds	X						X	
Carrots, angelica, anise, arracacha, asafoetida, caraway, carrot, celeriac, celery, centella asiatica, chervil, cicely, coriander/cilantro, cumin, dill, fennel, hemlock, lovage, Queen Anne's Lace, parsley, parsnip, and sea holly		X	X		X		X	
Butterhead, iceberg, loose leaf, and romaine lettuce	X						X	
Oranges, grapefruits, limes, and lemons	X				X		X	
Mixed fruit	X				X		XX	
Banana, abaca, and plantain		X	X		X			

Citations

"NPK-Value-of-Everything-Organic." AutoFlowerCulture, https://sites.google.com/site/autofloweculture/home/npk-value-of-everything-organic.

DAVENPORT, NIGEL. "NPK Value of Everything Organic!" The Nutrient Company, January 10, 2019, https://thenutrientcompany.com/blogs/horticulture/npk-value-of-everything-organic-database.

"COMPOSTING & PLANT NUTRITION." Compost and Composting, https://www.ibiblio.org/rge/course/compost.htm.

Table 21: Compost Checklist

Item	Yes	No	N/A
Are there goals for reducing food waste through vermicomposting?			
Are there goals for reducing yard waste through stockpile composting?			
Is stockpile compost rotated biweekly?			
Are there set goals for reducing food waste?			
Are there goals set for using compost teas as an alternative?			
Are carbon-nitrogen percentages from compost tested?			
Is there silt fencing around compost stockpiles?			
Is it incorporated into finished compost?			
Is compost incorporated into poor soil-texture areas?			
Score			

Compost Report Card

Good 7–9

Fair 4–6

Poor 1–3

Closing remarks

This book has been a twenty-year journey, with many failures and a couple of wins along the way with every challenge of our environment. We've explored the complexities of the natural world and the impact it has from stormwater runoff, bee hives collapse disorders and restoring native soils from harmful effects from excessive pesticide usage. This book has equipped you with the knowledge to understand the environmental issues we face. Now, it's your turn to take action. Please get involved, and advocate for change. Support sustainable practices, educate others, and hold decision-makers accountable. Remember, we are all stewards of this planet. The time for complacency is over. We must be the protectors of vulnerable ecosystems. Together, with informed action and unwavering commitment, we can create a brighter future for generations to come.